BULLETS & STEEL

BULLETS & STEEL

The Fight for The
GREAT KANAWHA VALLEY
1861-1865

Richard Andre, Stan Cohen & Bill Wintz

A War Time Scene in Buffalo, W. Va. 7th W. Va. Reg. of the U.S. Volunteers 1862.

The 8th WV Infantry at Buffalo, West Virginia.
See pages 30-31 for more information on this photograph.

quarrier press

Charleston, West Virginia

Library of Congress Catalog Card Number 95-72166

ISBN 13: 978-1-942294-29-0
ISBN 10: 1-942294-29-8

Originally published by Pictorial Histories Publishing Company.
Reprinted with permission by Quarrier Press.

Typography: Arrow Graphics
Layout: Stan Cohen
Cover Art: Richard Andre

The publisher would like to thank Terry Lowry for his invaluable assistance in this new printing.

Distributed by:

West Virginia Book Company
1125 Central Avenue
Charleston, West Virginia 25302
www.wvbookco.com

Preface

The main title of this book, "Bullets and Steel," comes from the tune sung by the Sandy Rangers of Wayne County during the Battle of Scary Creek. It rallied the Confederate troops to help turn the tide of battle.

This book covers the period of pre-Civil War, wartime and post-war history in the Great Kanawha Valley from Gauley Bridge in the east to Point Pleasant in the west.

It was a period of much turmoil in this border region adjacent to the Ohio River. Families were literally split apart by loyalties to both sides.

Gauley Bridge, at the foot of the imposing mountain barrier to the east and the river valley to the west, was a focal point for the conflict. Roads traversed the town both east to west and north to south. Charleston and the salt works to the east on the James River and Kanawha Turnpike were important military objectives in the control of the entire valley. Control would permit southern armies to threaten Ohio or northern armies to pressure the western part of Virginia. The salt works figured prominently in the strategies of both armies in the first two years of the war.

It is not the intention of the authors to present the definitive history of every event that occurred in the valley during the war. Instead, they want to give an overview of actions using as many first-person accounts as possible, as well as photos and drawings to enhance the story.

Several recent books have been written or edited by Terry Lowry, Tim McKinney, Bill Wintz and Dave Phillips that give a detailed account of certain aspects of the war in the valley. Historians such as Roy Bird Cook, Boyd Stutler and others have given good accounts of the actions.

The authors have used many different sources to give the reader a personal view of wartime history. Because of this there is some repetition of events, but from different perspectives.

Some of the material and photos are presented here for the first time! We are especially proud of the only known photo of the first capitol in Charleston.

This book could not have been done without the help of Terry Lowry, Tim McKinney and Paul Marshall, who provided articles and review of the manuscript. The staff of the West Virginia State Archives were most helpful with information and artifacts. Additional material was obtained from the West Virginia Regional History Collection, the Rutherford B. Hayes Library and the Virginia Military Institute Archives. Additional thanks for photographer Dan Davidson, typesetter Kitty Herrin, artists Mike Egeler, George McCausland and Julius E. Jones.

And thanks to the many other people who have contributed photos and information to the authors throughout the years.

Richard Andre, Stan Cohen
and Bill Wintz

Photo Identification

PHC—Pictorial Histories Collection
RBHL—Rutherford B. Hayes Library
WVSA—West Virginia State Archives
WVU—West Virginia Regional History Collection
USMHI—United States Military History Institute
Other photos are acknowledged to the source. Recent photos were taken by the authors.

Dedicated to the people of Kanawha Valley
past, present & future.

The Kanawha Valley in 1861

It can be said with certainty that "the winners write the history" and when considering the Civil War in the Kanawha Valley one should recall that for many years after the Confederate defeat historians were loathe to set down a truly balanced account of the conflict.

Perhaps the greatest of all Kanawha Valley historians, Dr. John P. Hale, himself once a Confederate artillery officer, gave the war but a very brief place in his accounts.

The fact is that in 1865 and for at least ten or fifteen years beyond it was the Yankees who called the tune and even what could be written and published.

By the turn of the century the South had recovered so much political power that candid recollections of old Confederate soldiers became more freely declared.

With this in mind let us consider the Kanawha Valley in April 1861 at the opening of hostilities.

The wealth of Charleston and Kanawha County was heavily involved with the labor-intensive salt industry and that labor was provided by slaves.

Salt furnaces dotted the riverbanks from above Malden to below Charleston. Salt, used to preserve foodstuffs, was of crucial importance to a Confederacy cut off from the only other sources in New York State.

Upon becoming president of the C.S.A., Jefferson Davis was faced with securing the Kanawha Valley for his new-born nation. From the beginning it was a lost cause due to the immense logistic problems in supplying Confederate armies across the Blue Ridge and Allegheny mountains. The Federals were able to steamboat up the Kanawha River from their stronghold in Ohio.

Some earlier writers have tried to depict Charleston as an unwilling, even disloyal, town that did not lend its support to the Confederate government in Richmond. The Kanawha Valley supplied soldiers for the C.S.A. and Col. George S. Patton's Kanawha Riflemen went on to gain fame as one of the most heroic companies in the Confederate army.

Imagine the feeling in Charleston as Fort Sumter was fired upon. You were a Virginian, there was no such thing as West Virginia, and Virginia extended to the Ohio River.

Like Robert E. Lee, you would consider yourself first a Virginian and then a citizen of the United States.

Perhaps your father owned a salt furnace and your home and livelihood depended upon it. The Federals proposed, indeed threatened, to invade your state and impose their will upon your society and community.

The point of view of young Lt. James C. Welch, who was to die in July 1861 at Scary Creek, was simply that he knew he had committed no crime—indeed he and his brothers wanted only to be left alone to pursue their peaceful and productive occupations.

While no defense can be made of slavery, nonetheless it was not forbidden under the U.S. Constitution.

The climate in the Kanawha Valley in April 1861 was one of fear. The fear that the Union army would soon occupy and overthrow the duly elected government.

It is perhaps difficult for us today, in the last decade of the 20th century, to understand just how our Kanawha Valley ancestors felt 134 years ago, but to sum up-they were by and large loyal to their native state of Virginia. They were determined to resist a hostile invasion by armed soldiers from the Unionist states and naturally they were heartbroken that events were inexorably leading to bloodshed.

Charleston had many ties to Ohio because of trade links and easy access by the river. There can be no doubt that residents of the Kanawha Valley had mixed emotions about taking up arms against what until a short while ago had been friends in Gallipolis, Pomeroy, Cincinnati.

By July the worst fears came true as the Union troops swarmed up the Kanawha and the smoldering fires of war burst forth on July 17, 1861, at the Battle of Scary Creek.

Inhumane acts of war were committed on both sides, as in every war, and when the Union army occupied Charleston it was noted that a group of drunken Kentucky infantry soldiers went into a small store demanding whiskey. The proprietor refused and the Kentuckians commenced to beat him whereupon the owner's son intervened in an attempt to save his father. In the ensuing fight a union soldier was shot in the leg.

For defending his father and his place of business, the young man was taken out on Kanawha Street and hanged!

Contents

Hearts Torn Asunder—Americans or Virginians?

When the question is asked: "Did Charleston sympathize with the North or the South in the Civil War?"; the answer cannot come until we state which Charleston—the Charleston of 1861 or the Charleston of 1865.

In the summer of 1861 Charleston was mostly loyal to Virginia and the Confederacy. The people of Charleston knew of the great Confederate victory at First Bull Run and it seemed clear the South would win the war. By 1862 it was less clear and by July 1863, after Gettysburg, it seemed the Confederacy was doomed. Human nature being what it is, we can safely say that enthusiasm for the Confederacy ebbed and flowed as battles were won or lost.

As late as the fall of 1862 the people of Charleston clung to the hope of final victory. This hope was strengthened on Sept. 13, 1862, when the hometown boys of the 22nd Virginia drove the Yankees headlong in retreat back to Ohio.

Spirits must have fallen only a month later when the Confederate army abandoned Charleston for the last time. To add to the demoralization of the local population was the grim day-to-day life under military occupation.

In July 1863 it is not hard to imagine the despair of most Charleston residents as news spread that General Lee was defeated and retreating from Gettysburg.

By 1864 most citizens were worn out and broken by the war. The brave resolve of 1861 gave way to the hollow-eyed exhaustion of 1865.

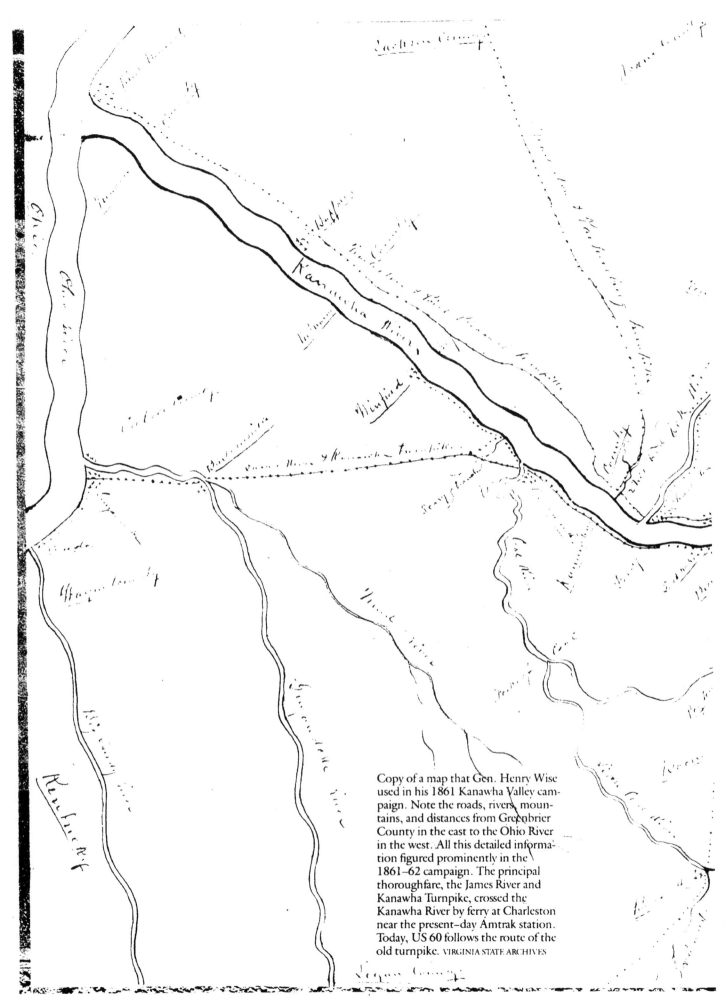

Copy of a map that Gen. Henry Wise
used in his 1861 Kanawha Valley campaign. Note the roads, rivers, mountains, and distances from Greenbrier
County in the east to the Ohio River
in the west. All this detailed information figured prominently in the
1861–62 campaign. The principal
thoroughfare, the James River and
Kanawha Turnpike, crossed the
Kanawha River by ferry at Charleston
near the present-day Amtrak station.
Today, US 60 follows the route of the
old turnpike. VIRGINIA STATE ARCHIVES

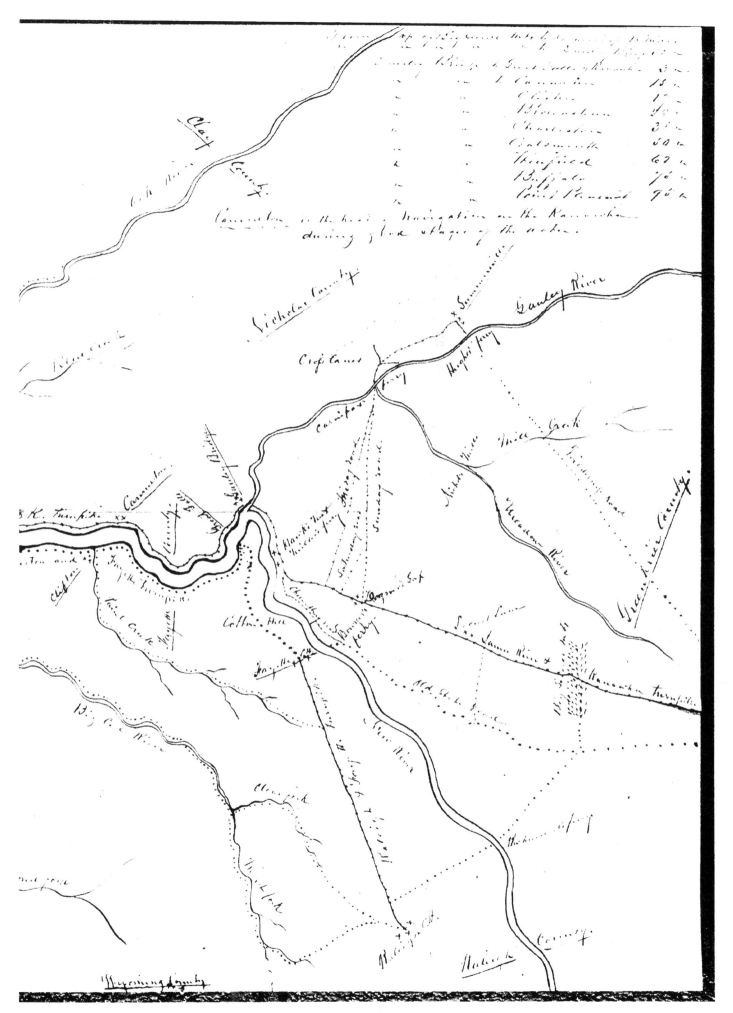

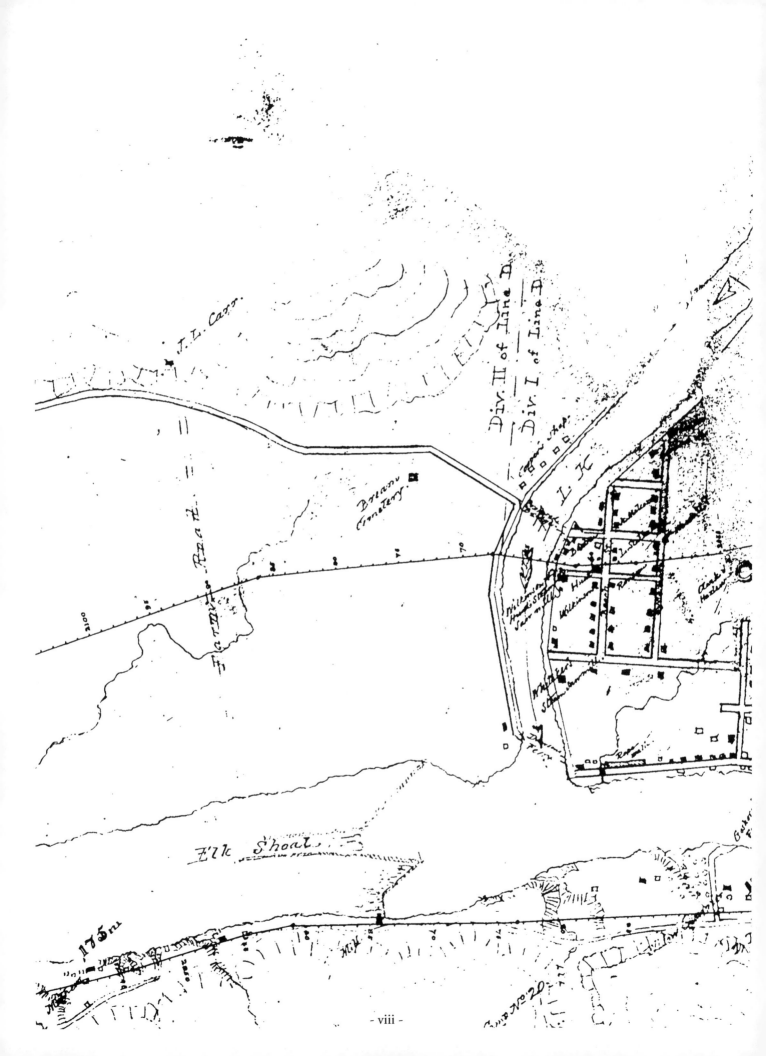

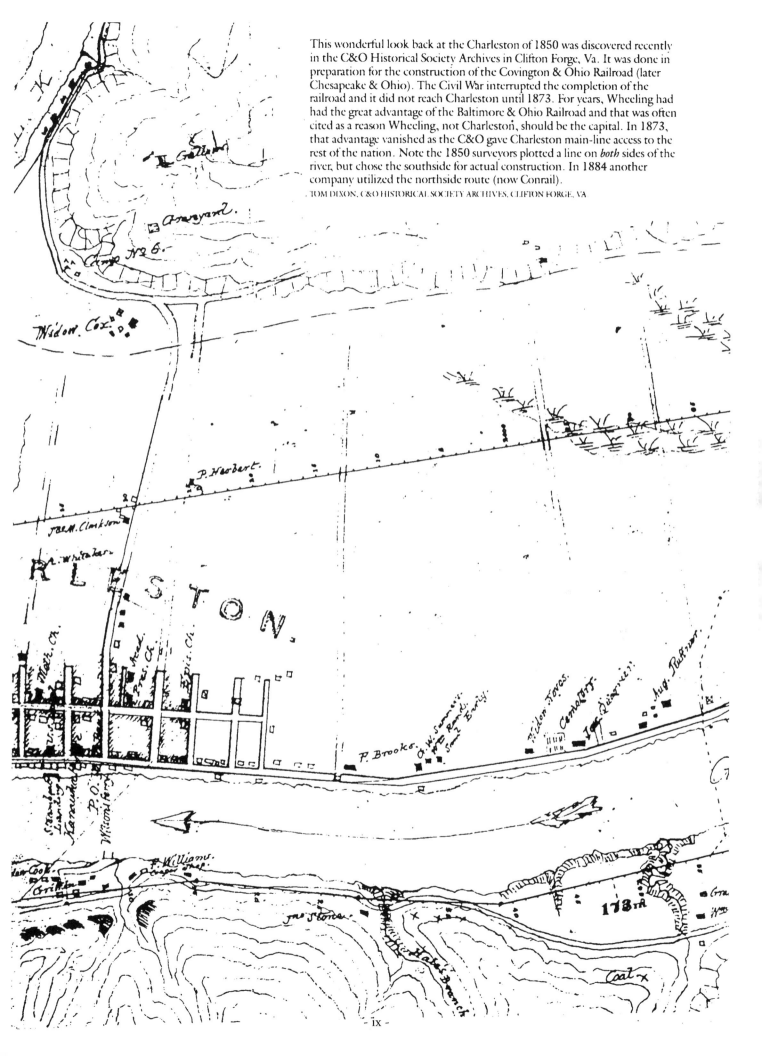

This wonderful look back at the Charleston of 1850 was discovered recently in the C&O Historical Society Archives in Clifton Forge, Va. It was done in preparation for the construction of the Covington & Ohio Railroad (later Chesapeake & Ohio). The Civil War interrupted the completion of the railroad and it did not reach Charleston until 1873. For years, Wheeling had had the great advantage of the Baltimore & Ohio Railroad and that was often cited as a reason Wheeling, not Charleston, should be the capital. In 1873, that advantage vanished as the C&O gave Charleston main-line access to the rest of the nation. Note the 1850 surveyors plotted a line on *both* sides of the river, but chose the southside for actual construction. In 1884 another company utilized the northside route (now Conrail).

TOM DIXON, C&O HISTORICAL SOCIETY ARCHIVES, CLIFTON FORGE, VA

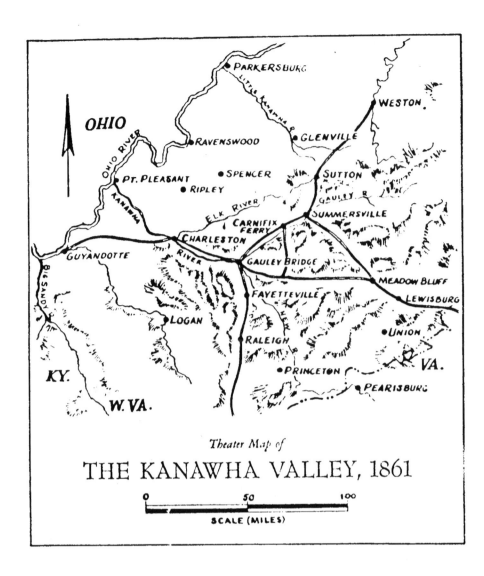

Theater Map of

THE KANAWHA VALLEY, 1861

```
0          50         100
```
SCALE (MILES)

The Division of Virginia in 1863 and the erection of West Virginia, has no parallel in history. The roots of this episode ran back into long years before the "War." The question of slavery was of minor importance. Indeed in all, forty-seven counties out of present West Virginia only had an average of two slaves to the square mile. But differences over commerce and education, the origin and habits of citizens, and Virginia's policy of internal improvements had caused to arise years before various schemes for division. At each constitutional convention able men from west of the mountains plead for a "fair deal." One governor alone had come from their number.

Nothing that could be written, however, no matter how fair the historian, would exactly suit the proponents of either side.

—Roy Bird Cook

A Pretty Little Town
The Pre-war Valley

Charleston in the 1860s

A map of Charleston drawn during the Civil War years shows the town having few streets and scattered buildings. As expected, it indicates a concentration of buildings along Front Street and Back (Virginia) Street and a number of buildings, presumably houses, in the Lovell Addition. Capitol Street, the focus of Charleston's existing downtown, is shown running from Front Street north all the way to the hill toward the northeast along Elk. The Cox mansion can be clearly seen just west of the road near the base of the hill. During the 1860s, Capitol Street was known as Cox's Lane, or the Cross Street leading to the road up Elk, or the road from Lewis D. Wilson's Steam Ferry. The latter name was used to identify the street until 1870, when the first capitol was built. The main reasons for earlier Capitol Street development were: it was part of the main transportation up the Elk River; and it was the terminus where a ferry operated across the Kanawha River, connecting the paths and roads on the south side with the paths and roads on the north side.

The map shows that very little development had occurred east of Broad Street and north of Donnally Street. That was because almost all of the land in those areas was either marshy or tied up in large farms, like the Cox plantation or the Ruffner farms. By the 1860s most of the original 1820s Cox holdings had been sold to John Clarkson, but some of the land remained in the Cox family until about 1871, when the heirs sold it to the Charleston Extension Company. Between present-day Quarrier Street and Donnally Street were the homesteads of the Quarriers, Dickinsons, Jeffries,

Whittekers, Burdettes and others. The land west of Capitol Street was owned by Thomas Fife, William Whitteker, Aaron Whitteker and Frank Noyes. From the 1880s until 1929 these properties were occupied by the Governor's Mansion, State Board of Control offices and later, the Daniel Boone Hotel, now 405 Capitol Street.

Most of the Lovell Addition was residential but the Civil War map does not differentiate between residential, business, and industrial. It is known that sawmills and a barrel manufacturing facility were located in the Lovell Addition vicinity after the war, but this survey did not confirm whether the businesses were there in the 1860s.

The block where City Hall stands today was in mixed use during the Civil War. John A. Truslow owned a store on Front Street. His building was a two-story brick structure. The building was occupied by a storehouse for T. Waggener & Company (no reference to the type of business). Orestes Wilson owned the rest of the lot and had a three-story brick building that housed his tavern and inn. The Virginia Street side of the block had two residences and a livery.

On the north side of Virginia Street between Alderson (Laidley) and Summers streets, the Methodists built a church in 1822 that came to be known as Asbury Chapel. The church was in the same location during the Civil War. The first Episcopal Church was located on the north side of Virginia Street at McFarland Street and the first Presbyterian Church was on the north side of Virginia at Hale, with the Mercer Academy behind it, facing where the Morrison Building now stands on Quarrier Street. Quarrier Street did not extend west of Hale Street at the time of the war. It is curious that Virginia Street was not known as Church Street in those days, as Capitol was later called Bank Street because of two banks on opposite corners of Capitol and Kanawha streets, and also at Virginia and Capitol streets.

There were many residences scattered through Charleston in the Civil War era. Most of them were small, vernacular buildings of little architectural merit, but there were a few fine architecturally stylistic buildings belonging to the more well-to-do citizens. Several of the outstanding ante-bellum houses are still standing today.

The Ruffner farms anchored the eastern end of today's Charleston beginning with the beautiful "Holly Grove" mansion erected by Daniel Ruffner in 1815, before the James River and Kanawha Turnpike (about where the Kanawha Boulevard is) passed by. Holly Grove was owned by Silas R. Ruffner during the Civil War. Not very far west of Holly Grove was the home of Joel Ruffner, called "Rosedale." The house was damaged during the Civil War, but survived only to be dismantled not too many years ago to make room for a modern building.

"Cedar Grove," the house built by Augustus Ruffner in 1834, still stands today at 1506 Kanawha Boulevard. It is a handsome brick farm house, well cared for and in very good condition. Augustus Ruffner was the son of Daniel Ruffner, builder of Holly Grove.

Another pre-Civil War house still standing in the east end of Charleston is the McFarland house, 1310 Kanawha Boulevard, built by Norris Whitteker, brother of Aaron, and sold to Henry Devol McFarland, son of merchant and banker James McFarland. During the Civil War years the house was owned by John C. Ruby II, who was a member of the Kanawha Riflemen. The house was used by the Federals as a hospital during the war. Just west of the Ruby house, during the war years, was the Rand house, also built by Norris Whitteker. It was razed in the 1940s.

Aaron Whitteker built the "Miller House" in about 1830 on Kanawha Street, just east of Broad Street. The house was originally owned by Thomas Friend, a salt-maker who also owned Friends' Vineyard and built the restored Dutch Hollow Wine Cellars in Dunbar. Friend sold the house to Samuel A. Miller in 1855. Mr. Miller was a major in the Confederate army. In the early days of World War II, the house was the headquarters for "Bundles for Britain." The building was razed in 1947.

Across Broad Street from the Miller house was the Andrew Donnally Jr. house, which was built at about the same time as Holly Grove (1815) and was among the earliest brick buildings in Charleston. The George Goshorn family lived in the house during the Civil War. Mr. Goshorn had come to Charleston in 1822 and first lived on the Kanawha River bank at the end of Goshorn Street, which was named for him. He operated the ferry at that point and also a hotel nearby. The house was demolished in 1942 to make room for a high-rise office building known as the United Carbon Building.

"Elm Grove," built by Rev. James Craik in about 1834, was located where the old Kanawha Valley Hospital stood. Actually, the house itself was near where Dunbar Street runs between Virginia and Quarrier streets. The lot was 154 feet wide along Virginia Street and incorporated an acre and a half.

Possibly the finest of the downtown Charleston houses was "The Elms," built by French architect John F. Faure in 1822. The house was known to Charlestonians as the "Brown House" because it was in the family of Judge James H. Brown from 1837 until it was demolished in 1959 to make room for a new federal building and the extension of Quarrier Street.

One of the finest and best preserved ante-bellum houses in Charleston is "Glenwood," on the west side. Built by Col. James Madison Laidley in 1852, the house belonged to George W. Summers II during the Civil War. The house today belongs to the West Virginia College of Graduate Studies Foundation, which administers the mansion and its grounds.

Glenwood was one of five farms fronting the Parkersburg Pike before and during the Civil War. Edgewood, built in 1848 by J. L. Carr, was located on a bluff just east of the drive that today bears the estate's name. William Gillison owned the next farm west of Glenwood. Farther west at a point that would be on today's Beech Avenue above the north end of Patrick Street, Dr. Spicer Patrick built the house in 1855. The westernmost farm was Littlepage, with its stone mansion built on bottom land near Kanawha Two Mile Creek. The house was built in 1845 and during the Civil War its grounds served as camping ground for both Federal and Confederate troops at various times beginning with Wise's Legion in 1861.

*From *Fort Scammon and A History Of The Civil War In Charleston And The Kanawha Valley, West Virginia*, Paul D. Marshall & Assoc. Inc., 1986

In 1859 Virginians Draft Resolve of Their State's Rights

A meeting was held in the Kanawha County courthouse on Dec. 19, 1859. The *Kanawha Valley Star* reported on it in its Dec. 26 edition. The article showed the deep concern for Southern rights prevalent in the years before the outbreak of civil war.

According to the article, the meeting was attended "by the oldest, most eminent, and conservative men of the County." Judge George Summers was elected chairman and A. W. Quarrier, secretary. Benjamin H. Smith, Dr. Spicer Patrick, Thomas L. Broun, John D. Lewis, John S. Swann, James M. Laidley, James H. Fry, Jacob Goshorn and Nicholas Fitzhugh were appointed to draft resolutions of the meeting.

The assembly was addressed by George S. Patton, Dr. John Parks and James H. Brown. Benjamin Smith then came forward with the committee's draft of the Resolves. The preamble declared that:

Virginia, as one of the Sovereign States of the Union has Always conformed to, and strictly complied with, her obligations under the Constitution; whilst the history of some of the nonslave holding States for the last thirty years, and particularly of late years, has been a series of encroachments upon the violations of the Constitutional Rights of Slaveholding States; and the recent events at Harper's Ferry, and the sympathy expressed from the Pulpit, by the Press, and by the People of the North for John Brown and his fanatical associates, have assured us of the South that a large portion of the Northern people are ready and willing to do any and everything that will overthrow the domestic institutions of the South; and that the endorsement of Helper's infamous book by sixty-eight Black Republican Congressmen, by the Governor of New York, by ex-Judges and other prominent persons of the North, plainly indicates a deadly hostility and bitter hatred on the part of the Black Republicans towards the South, in utter disregard of the Constitution of the United States, and of the Common Rights of Humanity."

The *Resolves* started with a statement of approbation of "the dignity and decorum, as well as the stern justice, exhibited by Virginia and her people; and commended the courage and wisdom displayed by Governor Wise all the way through the period of tension and excitement. In language clear and strong was expressed "the willingness of KANAWHA to perform her part in effecting any measures that Virginia and her sister Southern States might deem necessary for the protection of their Rights, etc."

That the county of Kanawha would "sanction and approve all retaliatory measures against the non-slaveholding States which the wisdom of the General Assembly may see fit to enact."

Judge Summers had a remarkably incisive mind and a keen perception of issues. His concluding opinion was to the effect that the Union could not be "advantageously preserved" by force. Even prominent newspaperman Horace Greeley declared "Let the erring sisters go." Summers closing remarks were eloquent:

I have lived all my days in Virginia, and have rarely been beyond her limits. I love her as a son loves his mother. My first duty is to her, and where she goes, by the solemn judgement of her people, I go. Her destiny is my destiny. But I desire, also, to retain the name and the privileges of an American citizen, and am unwilling to give them up, until every possible effort has been fully made for their preservation. God grant that our institutions, both State and National, may come out of this fiery trial unharmed.

With great respect and esteem,
Yours very truly,
Geo. W. Summers.

The Ruffner Pamphlet

No history of the Civil War in the Kanawha Valley would be complete without recalling the 1847 anti-slavery pamphlet written by Henry Ruffner some 14 years before the war.

Henry Ruffner was president of Washington College in Lexington, Va. (later Washington & Lee). He was the son of David Ruffner and was very familiar with the institution of slavery since he had lived in Kanawha Salines much of his life.

The Ruffner pamphlet engendered much debate throughout Virginia and as its author Ruffner was subjected to harsh insults and derision for his controversial view.

Although Ruffner certainly condemned slavery for moral reasons he primarily attacked it on economic grounds. He correctly pointed out that the institution of slavery was very detrimental to the development of Virginia because free white workers and craftsmen could not compete with slave labor. The result being a constant migration of native white Virginians to other states where slavery was not practiced.

There is little doubt that wage-earning whites were at a great disadvantage trying to earn a living in areas where slave labor existed. Ruffner had seen this first-hand in the salt works of the Kanawha Valley, where it was well known that the salt makers could not turn a profit if required to hire free men at market wages.

As competition from other salt sources became more intense, so did the need for slave labor to continue production.

Much to his credit, Ruffner was more interested in the long-term development of the entire state than the future of the Kanawha Valley salt industry. What he dared to write was already well known and agreed upon by many western Virginians, very few of whom owned slaves outside of the Kanawha Valley.

What must be understood, however, is that those who took up arms for the Confederacy did so not to uphold slavery but to oppose invasion by Federal armies and what they considered the unconstitutional violation of Virginia's sovereign state's rights.

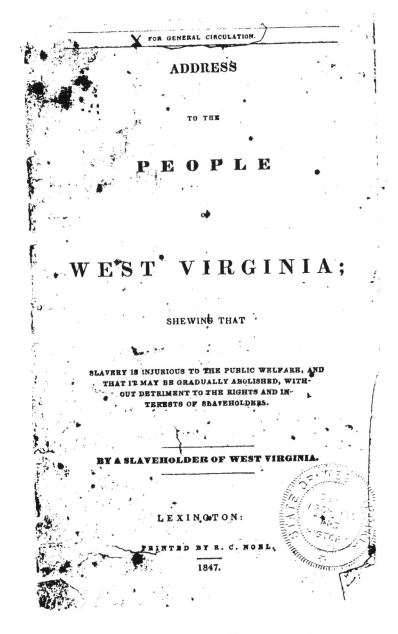

FOR GENERAL CIRCULATION.

ADDRESS

TO THE

PEOPLE

of

WEST VIRGINIA;

SHEWING THAT

SLAVERY IS INJURIOUS TO THE PUBLIC WELFARE, AND THAT IT MAY BE GRADUALLY ABOLISHED, WITHOUT DETRIMENT TO THE RIGHTS AND INTERESTS OF SLAVEHOLDERS.

BY A SLAVEHOLDER OF WEST VIRGINIA.

LEXINGTON:

PRINTED BY R. C. NOEL,

1847.

Henry Ruffner was born in the Shenandoah Valley in 1790. He was a preacher in the Southern Presbyterian Church in Charleston and was president of Washington College. He married Sarah Montgomery Lyle of Rockbridge County, Va., and they later settled in Kanawha Salines. He died in December 1861.

until the close of the important debate in the Franklin Society, to which your letter alludes. The arguments delivered by several of yourselves, and the results of my own examination of facts, so impressed my mind with the importance of the subject to the welfare of the country, that I proceeded immediately to write out an argument in favor of a gradual removal of slavery from my native soil, our dear West Virginia; and intended in some way to present it to the consideration of my fellow-citizens. Some months ago you privately signified a desire that it might be printed, and have now formally made the request.

I cheerfully comply, so far as this, in the first instance, that I will prepare for the press an Address to the Citizens of West Virginia, comprising the substance of the argument as delivered by me, enriched and strengthened by some of the impressive views exhibited by several of yourselves. Within the limits of a moderately sized pamphlet, it is impossible to introduce every important consideration bearing on the subject, or to do more than present the substance of the prominent facts and reasons which were more fully exhibited and illustrated by the debaters in the Society.

As we are nearly all slaveholders, and none of us approve of the principles and measures of the sect of abolitionists, we think that no man can be offended with us for offering to the people an argument, whose sole object is to show that the prosperity of our West Virginia —if not of East Virginia also,—would be promoted by removing gradually the institution of slavery, in a manner consistent with the rights and interests of slaveholders.

To the Great Being who rules the destinies of our country, I commit the issue of this important movement.

Yours,

HENRY RUFFNER.

CORRESPONDENCE.

LEXINGTON, VA., SEPT. 1st, 1847.

Dear Sir:

The undersigned believing that the argument recently delivered by you in the Franklin Society, in favor of the removal of the negro population from Western Virginia, was not only able but unanswerable; and that its publication will tend to bring the public mind to a correct conclusion on that momentous question; request that you will furnish us with a full statement of that argument for the press.

We cannot expect that you will now be able to furnish us with the speech precisely as it was delivered, nor is it our wish that you shall confine yourself strictly to the views then expressed. Our desire is to have the whole argument in favor of the proposition, presented to the public, in a perspicuous and condensed form. And believing that your views were not only forcible but conclusive, and that they were presented in a shape, which cannot give just cause of offence to even those who are most fastidious and excitable on all subjects having any connexion with the subject of slavery, we trust that you will be disposed cheerfully to comply with our request above expressed.

Very Respectfully,
Your ob't serv'ts,
S. McD. MOORE,
JOHN LETCHER,
DAVID P. CURRY,
JAMES G. HAMILTON,
GEORGE A. BAKER,
J. H. LACY,
JOHN ECHOLS,
JAMES R. JORDAN,
JACOB FULLER, Jr.,
D. E. MOORE,
JOHN W. FULLER.

The Rev. HENRY RUFFNER, D. D.

LEXINGTON, VA., September 4th, 1847.

To Messrs. Moore, Letcher &c.,
GENTLEMEN:

Though long opposed in feeling to the perpetuation of slavery, yet like others I felt no call to immediate action to promote its removal,

ADDRESS

TO THE

CITIZENS OF WEST VIRGINIA.

FELLOW-CITIZENS,

Now is the time, when we of West Virginia should review our public affairs, and consider what measures are necessary and expedient to promote the welfare of ourselves and our posterity. Three years hence another census of the United States will have been completed. Then it will appear how large a majority we are of the citizens of this commonwealth, and how unjust it is that our fellow citizens of East Virginia, being a minority of the people, should be able, by means of their majority in the Legislature, to govern both East and West for their own advantage. You have striven in vain to get this inequality of representation rectified. The same legislative majority has used the power of which we complain, to make all our complaints fruitless, and to retain the ascendancy now when they represent a minority of the people, which they secured to themselves eighteen years ago, while they yet represented the majority.

You have submitted patiently, heretofore, to the refusal of the East to let West Virginia grow in political power as she has grown in population and wealth. Though you will not cease to urge your claims, you will, if necessary, still exercise this patient forbearance, until the next census shall furnish you with an argument, which cannot be resisted with any show of reason. Then—as it seems to be understood among us—you will make a final and decisive effort to obtain your just weight in the government.

That will be a critical period in your public affairs. A great end will then be gained, or a great failure will be experienced. Are you sure of success? Can you be sure of it, while the question of representation stands alone, and liable to unpropitious influences, even on our side of the Blue Ridge? We propose to strengthen this cause, by connecting with it another of equally momentous consequence—in some respects even more—to our public welfare. United they will stand; divided they may fall.

You claim the white basis of representation, on the republican

This pamphlet provides very interesting reading, written 14 years before the beginning of the Civil War. Notice that Ruffner addresses the citizens of West Virginia not citizens of western Virginia. West Virginia did not become a state until 1863.

The Kanawha County Courthouse was built in 1817 and used until the present building was constructed on the same site in 1892. Both sides used this site when they occupied Charleston. WVSA

Modern view of remodeled Elk River house · Lt. Gen. Jubal A. Early, C.S.A.

Jubal A. Early's Kanawha Valley Connection

George W. Summers wrote in a 1939 *Charleston Daily Mail* article that Confederate General Jubal Early lived on the Elk River above Charleston for a period during the 1840s. Early was born in Franklin County, Virginia, graduated from West Point, and practiced law in Franklin County from 1840 to 1846.

For some unknown reason, perhaps a love affair gone bad, he abandoned his law career in the 1840s and supposedly moved to the Kanawha Valley, where he lived as a recluse.

The authenticity of Early's Kanawha Valley connection has not been substantiated. However it seems that his father, after becoming a widower in 1832, married a widow whose last name was Cabell. In the Kanawha County Deed Book E, page 204, dated 1819 is a deed under the name John J. Cabell for 500 acres, five miles from the mouth of the Elk River on the north side—the exact location of the cabin that Early supposedly built in the 1840s. (It is documented that he spent two months with his father in Buffalo, Putnam County, in the fall of 1847.)

General Early fought under Andrew Jackson in the Seminole Indian campaign in Florida and the Mexican War. In the Civil War he was considered one of the ablest generals of the Confederate Army. In 1864 Early was given a large and completely independent command in the Shenandoah Valley, where his assignment was to draw Union forces away from Richmond. Faced with little opposition, he swept down the valley and was preparing for an attack on Washington itself. Early's advance troops entered the District of Columbia; within sight of the Capitol, an action viewed by Lincoln. Two corps having been dispatched by Grant pushed Early back.

Early was leading his troops to victory at Cedar Creek when Gen. Philip Sheridan made his famous "20-mile ride," from Winchester. In March 1865, his forces were virtually destroyed by Sheridan at Waynesboro.

Early never returned to his pre-war property in the Kanawha Valley, but the still-standing cabin called "Viking Hill," formerly an inn, has been remodeled and expanded into apartments.

The Salt Industry

Charleston, Virginia, was 320 miles and across two great mountain ranges from the capitol in Richmond. Indeed, Charleston was so near Ohio that slaves hired out from the Tidewater plantations to the salt works along the Kanawha brought higher rental fees because of the danger they might escape to freedom.

The lower Kanawha Valley consisted of rich productive farmlands, but the true money crop of Charleston was salt, and salt was produced by slave labor. By the 1860s competitive salt-producing areas pressured profit margins among the Kanawha salt makers and it was said that the wages necessary to pay white laborers would have ended the industry in the Kanawha Valley.

Dr. John P. Hale and his cohorts had brought the level of drilling and distilling technology to a zenith and in truth, regardless of the Civil War, the handwriting was on the wall for Kanawha Valley salt makers as new areas opened up across the nation.

From 1800 up until the 1850s great fortunes were made in the salt business and it was in many ways akin to gold mining. The growing nation had an inexhaustible need for salt and the Kanawha Valley salt furnaces produced it by the millions of barrels. Unlike the rest of western Virginia with its mountaineer farmers, the slave labor that drove the industry was more like the deep South cotton plantations. By 1861 the salt business was in decline and the destruction of war sped that decline.

Dr. John P. Hale, a leading citizen of Charleston, was a key figure in two important elements of the city's development. Prior to the Civil War, Hale owned and operated the biggest salt business in the nation and after the war he was most responsible for seeing that the capitol was built in Charleston.

———————— ✦ ————————

"Kanawha Salines is a flourishing village 6 miles above Charleston. The salt works on the Kanawha are very extensive, employing near 3000 persons in their operations. Near three million bushels of salt is here manufactured annually, which find a ready market in the States of Ohio, Indiana, Illinois, Missouri, Tennessee and Kentucky."

ELLIOTT & NYE'S VIRGINIA DIRECTORY, 1852

View of the Salt-Works on the Kanawha.

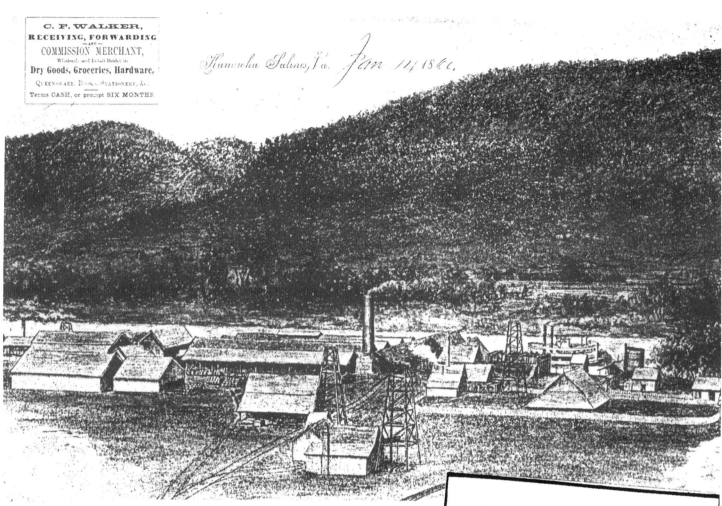

Kanawha Salines, Va. Jan 14, 1864.

This pen-and-ink drawing of the Dickinson Salt Works at the east end of Malden was done about 1898. However, it can be assumed that not much had changed in the 35 years since the Civil War. The wells are obvious as is the furnace. A steamboat can be seen loading up for the trip to the great meat packing city of Cincinnati, humorously called "Porkopolis."

In 1864 the president of Kanawha Salt Company, John P. Hale, and several other pre-war salt makers attempted to reorganize the salt cartel to control the price and distribution of local salt. By this time, however, the salt industry was devastated by competition of western salt producers and the ravages of Civil War. The new Kanawha Salt Company was formed during the Union army occupation of the Kanawha Valley, although its president, John P. Hale, had joined the Confederate army in 1861. It has not been determined how Hale managed to put this company together or to whom salt was sold during the war, but by the time the company was organized the heyday for the salt industry was over in Kanawha County.

KANAWHA SALINES, January 23, 1864.

At a full meeting of the members of the Kanawha Salt Company on the 21st inst., it was

Resolved, That Dr. John P. Hale, President of the Kanawha Salt Company, be authorized to sign the name of said Company in its legitimate transactions, in accordance with the provisions of their constitution.

J. P. HALE, *President.*
N. B. COLEMAN, *Secretary.*

KANAWHA SALT COMPANY.

Dr. J. P. HALE,
President and Office Agent.

GEN. L. RUFFNER,
General Advisory Agent.

J. D. LEWIS,
A. P. FRY,
Consulting and Shipping Agents.

MORRISON & OAKES,
A. P. FRY,
LEWIS & SON,
WALKER & SHREWSBURY,
H. W. REYNOLDS,

H. H. WOOD,
F. A. LAIDLEY,
JAMES H. FRY,
L. RUFFNER, JR.,
J. E. THAYER & CO.,
Directors.

CINCINNATI:
Moore, Wilstach & Baldwin, Printers,
25 WEST FOURTH STREET.
1864.

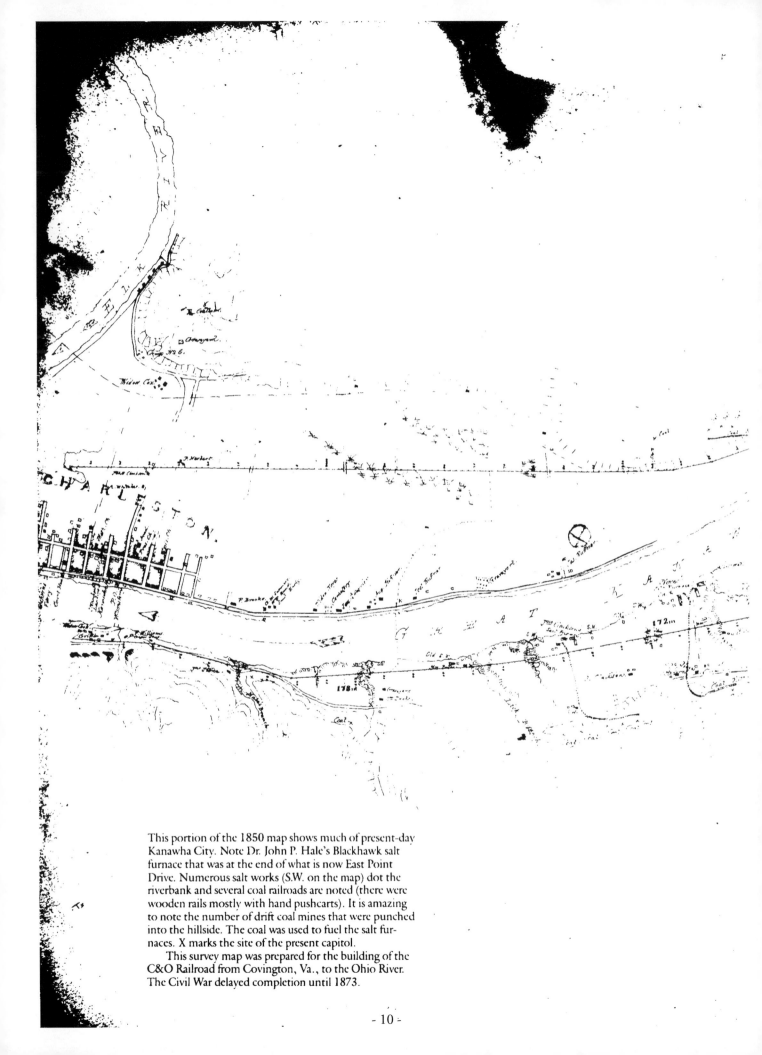

This portion of the 1850 map shows much of present-day
Kanawha City. Note Dr. John P. Hale's Blackhawk salt
furnace that was at the end of what is now East Point
Drive. Numerous salt works (S.W. on the map) dot the
riverbank and several coal railroads are noted (there were
wooden rails mostly with hand pushcarts). It is amazing
to note the number of drift coal mines that were punched
into the hillside. The coal was used to fuel the salt furnaces. X marks the site of the present capitol.
    This survey map was prepared for the building of the
C&O Railroad from Covington, Va., to the Ohio River.
The Civil War delayed completion until 1873.

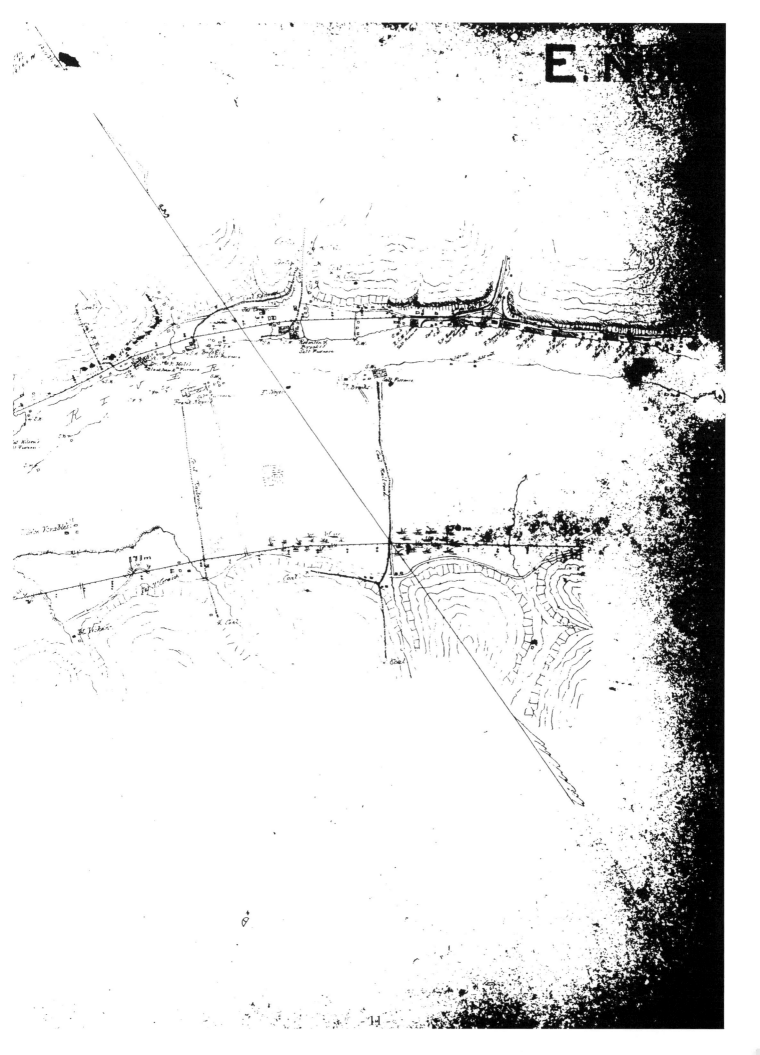

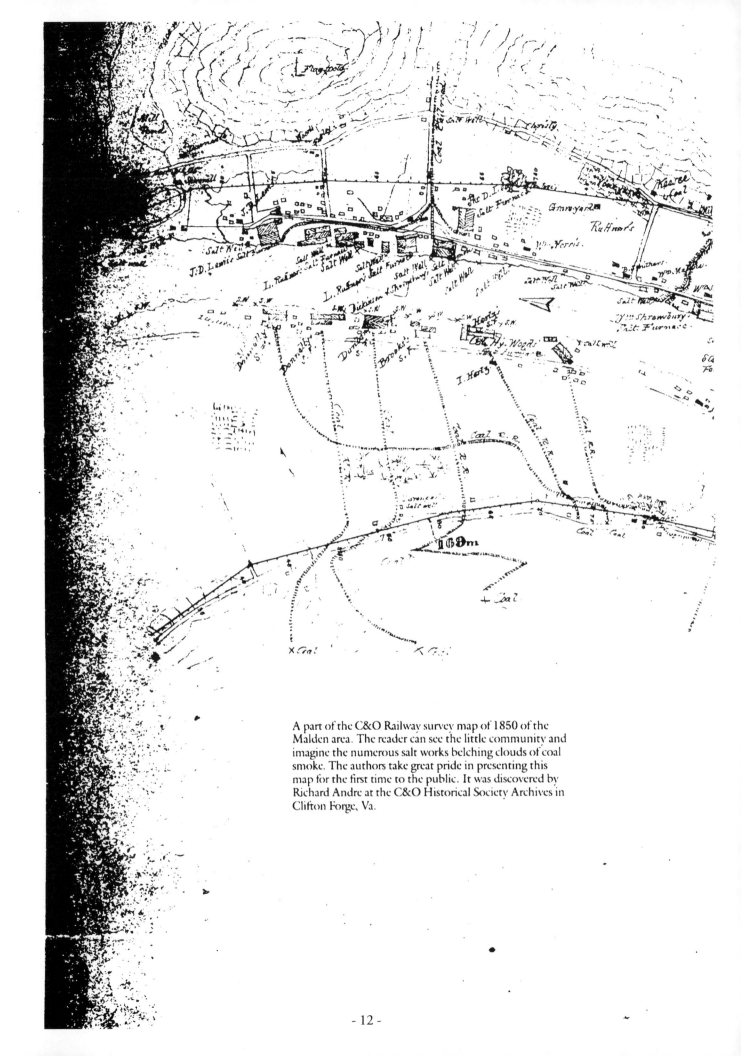

A part of the C&O Railway survey map of 1850 of the Malden area. The reader can see the little community and imagine the numerous salt works belching clouds of coal smoke. The authors take great pride in presenting this map for the first time to the public. It was discovered by Richard Andre at the C&O Historical Society Archives in Clifton Forge, Va.

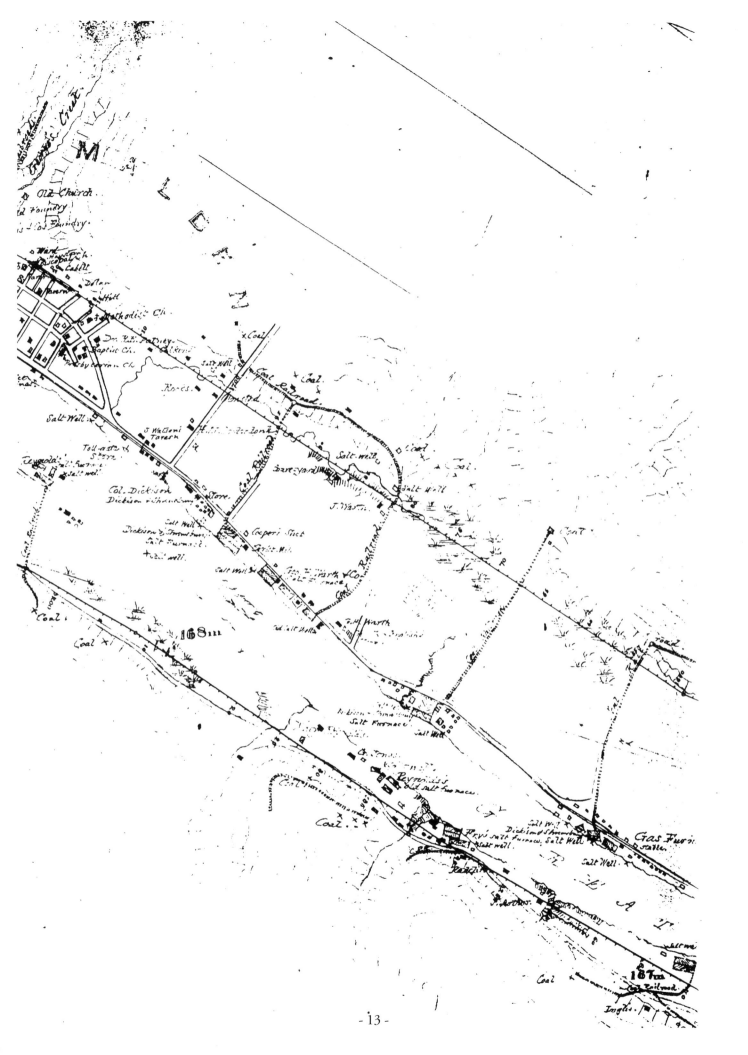

1842 postal stamp,
Kanawha Salines, Virginia.

The Kanawha Salines
Presbyterian Church is the
oldest church in Malden,
and one of the earliest
Presbyterian churches in the
county. The congregation
was formed in Charleston in
1819, and this structure was
completed in 1840. In
1933, a Sunday school
building was added at the
rear of the church. Henry
Ruffner was one of the
preachers at the church. The
church was integrated and a
gallery was provided for the
slaves of the salt producers.

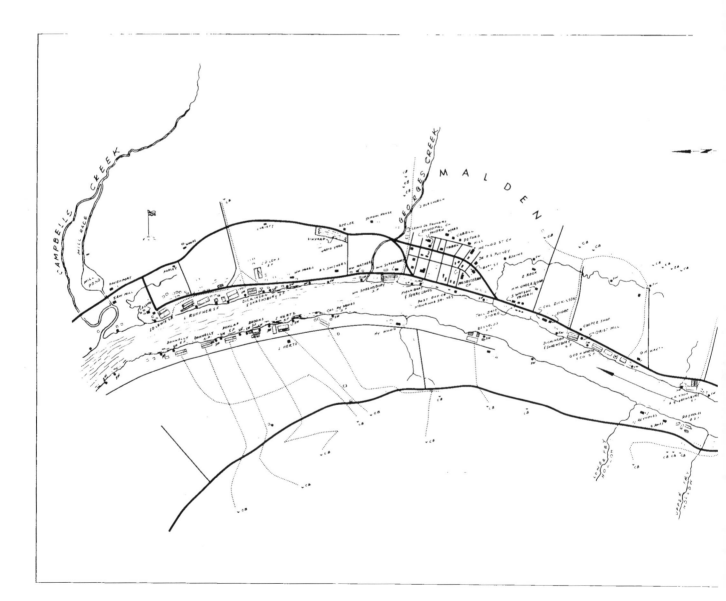

The Richard E. Putney house is one of the most imposing and least-altered homes in Malden. The principal two-story brick structure was built about 1836. The one-story addition was erected in at least two stages and the wooden front porch is also a later addition. The first owner, Dr. Richard E. Putney, married the daughter of David Ruffner. In 1868 the Kanawha Salines Presbyterian Church bought the house for use as a manse. It was a private home from 1952 until 1973—when it was converted into a law office.

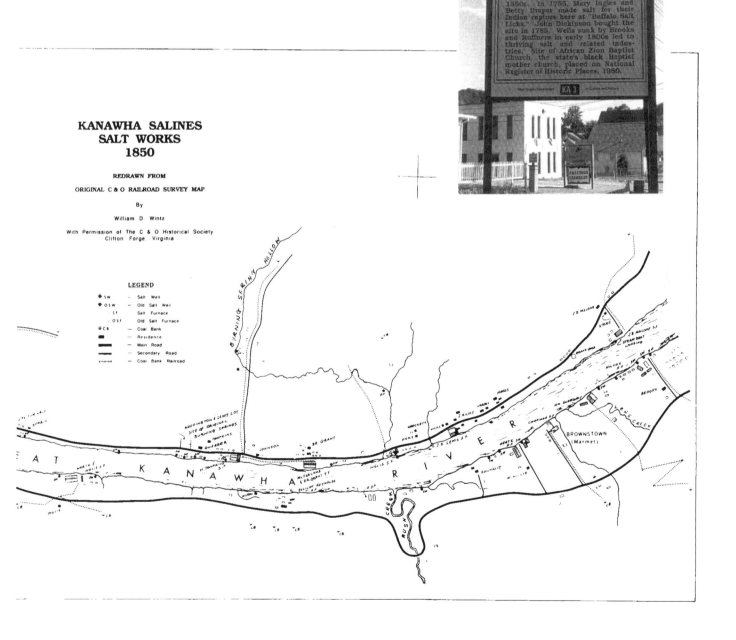

KANAWHA SALINES
SALT WORKS
1850

REDRAWN FROM

ORIGINAL C & O RAILROAD SURVEY MAP

By

William D Wintz

With Permission of The C & O Historical Society
Clifton Forge Virginia

LEGEND

◆ SW — Salt Well
◆ OSW — Old Salt Well
SF — Salt Furnace
OSF — Old Salt Furnace
CB — Coal Bank
— Residence
— Main Road
— Secondary Road
— Coal Bank Railroad

This painting of the Charleston riverfront was done in 1854 and appeared in Parish's *Art Work of the Kanawha and New River Valleys, 1897*. The town had about 1,500 people and a number of buildings, churches and houses clustered in the current downtown area. In the center of the painting is the present-day levee. WVU

A 10-pound smooth-bone iron cannon was cast at a foundry in Malden by O. A. and W. T. Thayer. It was used in the Scary battle and nicknamed "Peacemaker." When they retook the valley they went to Malden to find the wooden patterns for this weapon. Not finding it they arrested one of the Thayers and threatened to send him to prison. The pattern had been secretly sent to an old woman in the Campbell's Creek area, who hid it in her bed. The cannon was soon captured by the Federal troops.

Charleston's first mayor, Jacob Goshorn, was elected only four months before Federal troops moved into Charleston following the Battle of Scary Creek in July 1861. Goshorn surrendered the town to the Federal troops, then fled south for the remainder of the war. He returned after the war to resume his business career.
OFFICE OF THE MAYOR, CHARLESTON

Their Native State
Military Units

The Kanawha Riflemen

As the Kanawha Valley lay entirely on Virginia soil, it saw the immediate formation of a number of state and militia volunteers, organized for defense against any invaders or aggressors. In fact, the actual nucleus of the Southern army in the Kanawha Valley, as well as the 22nd Virginia in particular, originated through the foresight of Capt. George Smith Patton, grandfather of Gen. George S. Patton of World War II fame. Patton, an 1852 graduate of the Virginia Military Institute, had spent time studying law in his father's Richmond office. During that time he had often viewed nearby the drilling of the Richmond Light Infantry Blues, Virginia's oldest and among its most distinguished military organizations.

In 1856 Patton and his new bride moved to Charleston in the Kanawha Valley where he set up a law practice with Thomas Lee Broun. In his new profession he served as Commissioner of Chancery to Kanawha Circuit Court and Kanawha County Court. Ironically, Patton and Broun both served on the board of directors of the Coal River Navigation Company while William Starke Rosecrans, a future military opponent, was company president.

Upon arrival in Charleston, principal economic center and largest town in the Kanawha Valley, Patton purchased the "Elm Grove" estate and quickly became a popular citizen, nicknamed "Frenchy" due to his pointed mustache. He was self-confident, a smart dresser, and displayed classic chivalry toward ladies, making him a dashing, romantic figure. Patton undoubtedly anticipated the forthcoming conflict and formed a militia company in 1856, dubbing it the Kanawha Minutemen. Their formation took place at Brooks Hall (also known as the "Assembly Room") in Charleston, and about a week later they elected officers. The name soon changed to the Kanawha Rifles, and finally, in November of 1859, it changed permanently to the Kanawha Riflemen.

Charleston's newspaper, the *Kanawha Valley Star*, reported of the Riflemen: "This Company is already large and rapidly increasing. It is holding drills regularly." In fact, according to a veteran of the unit, the earliest organization was a "drilling club," something of which Patton was a master. But another statement in the by-laws hinted at their seriousness:

> No provision is made for the failure of a member of this company to attend in time of war or tumult, for it is naturally supposed that every member will be present. However, should any member fail to answer roll call under such circumstances, and not show good reason therefor, he shall be dealt with as a deserter from the army.

The Kanawha Riflemen, with Patton as captain, obviously was patterned after the Richmond Light Infantry Blues, and was composed of the socially prominent and wealthy men of the valley, including some 20 lawyers. The unit was privately financed. Patton, a sharp disciplinarian, drilled the men extensively in the old Mercer Military School lot, in Joel Ruffner's meadow (present Ruffner Park, where a commemorative marker to the Riflemen now rests), and in Brooks Hall. Because they were impressive in drill, the Riflemen were often invited to appear at parades, balls, and social functions, earning a reputation that they could dance as well as, and maybe better than, they could fight. Their decimated ranks at the end of the war put that belief to rest forever.

When the Riflemen appeared at such social affairs, Patton had the guard mounted. As one of them said, "I recall a July 4th celebration and picnic at Ruffner grove, in Slab Town as we used to call the eastern end of the city, and the Riflemen turned out en masse. I was one of the mounted guards, marching back and forth for hours in the hot sun, in my new uniform with my rifle on my shoulder—proudly guarding the fried chicken and lemonade."

This life of leisure turned serious in 1859 when there was talk of sending the Riflemen to Harpers Ferry to help quiet the John Brown affair, but this never materialized. However, in the autumn of 1860 another incident almost erupted into violence when the Riflemen were invited to attend the Agricultural Fair at Point Pleasant. Two uniformed members took occasion to visit friends at nearby Gallipolis, Ohio, and soon found themselves jeered and harassed by the strongly pro-Union townsfolk. They were quickly rescued by a small detachment of the Riflemen. According to Noyes Rand, one of the unwelcome visitors (and a future adjutant of the 22nd Virginia), had it not been for the discipline under which the Riflemen had been so well trained by Patton, tempers would have flared and "hostilities would have begun then & there in Gallipolis instead of next Spring at Ft. Sumter."

The Kanawha Riflemen included a fluctuating membership of some 75 to 100 men, although member Joseph A. Brown said there were 120. The Riflemen had an excellent brass band led by an English cornetist. They were best known for their gaudy uniforms which Patton had personally designed for them. Slightly varying descriptions survive, but the uniform basically consisted of brass buttons stamped with the Virginia coat of arms on a uniform of "dark [olive] green broadcloth, matching overcoats, a fancy headgear consisting of black hats with ostrich feathers dangling from the wide brim with a gold KR on the front of the cap, the entire headdress covered with oilcloth in bad weather, and white [cotton] Berlin gloves." The "coats were long, with epaulets of gold braid . . . and a short shoulder cape graced the coat." Patton also required the men to blacken their shoes regularly. This uniform was undoubtedly patterned after that of the Richmond Light Infantry Blues. When the Riflemen later mustered into the Confederate army as Company H, 22nd Virginia Infantry, authorities ordered them to dispose of the flashy

uniforms. The new more modest outfit featured "light blue jackets and dark gray trousers with yellow trimmings." Some men still wore the original uniform in the early campaigns of 1861.

Weaponry of the Riflemen consisted of a various assortment of guns, including "a short Neiss Rifle, with bayonet, cartridge box and scabbard." Others carried Belgian rifles or a Harpers Ferry Rifle which had a bore of about .45, an iron ramrod, fired with percussion caps, and had a grease box located in the stock. Joseph A. Brown said that they were armed with the noted Mississippi Rifle. One Rifleman reflected:

Although there is no record of it, I think Patton must have offered the services of the company as a whole to the Governor very shortly after we organized. For had we not been part of the Virginia state militia, and under the command of old Colonel Hogue of Winfield, commandant of this district, we could not have secured rifles from the state. We were inordinately proud of our appearance. After we had learned to drill we never lost an opportunity to turn out, especially after the rifles had come from Richmond. I recall they were locked in Brooks Hall under guard and none of us were allowed to look at those precious rifles at first. I managed to get a peep at them, however, by calling through the door that Colonel Hogue had come and wanted to inspect the guns. When the door was opened I walked in.

The proud ladies of Charleston bought the Kanawha Riflemen a flag, which had the regular 13 stripes and the Virginia coat of arms. With the outbreak of the war they converted the stripes to the Confederate colors and left the coat of arms.

So it was that Patton, the transplanted Richmond lawyer, full of daring and dash and only 24 years of age in 1861, prepared the Kanawha Valley of western Virginia for its defense well before the war began. Patton was not alone. A number of other militia groups organized in the valley prior to the war, the most prominent being the Kanawha (or Charleston) Sharpshooters. This company was composed of many Kanawha River boatmen and was organized in 1859 by Captain John Sterling Swann, a Charleston lawyer. He was assisted by Joel Ruffner Jr., 1st Lieutenant John R. Taylor, 2nd Lieutenant Charles Ufferman, 3rd Lieutenant Dr. A. E. Summers, and Sergeant James Venable. Although some sources claim that the Sharpshooters wore gray uniforms, according to the minutes of their meetings on March 12, 1860, they adopted a uniform of navy blue flannel material. It included a high-collared frock coat with a single row of nine buttons, a short shoulder cape, and pants with two green cords at the seams. It is known that in 1860 the Sharpshooters received permission to drill in a field north of Charleston belonging to a Mr. Clarkson, and that later they were often found practicing maneuvers in a field belonging to the Welch family. The Sharpshooters often worked in conjunction with the Kanawha Riflemen. Once, when both companies attended the funeral of a kinsman of General Winfield Scott, the *Kanawha Valley Star* said: "The two companies were handsomely uniformed and equipped, presenting an imposing appearance. Captain Swann's Company has exhibited its heroism frequently in the capacity of a fire brigade, battling at all hours of the night in heat and cold." The Charleston Sharpshooters would serve only briefly with the 22nd Virginia Infantry, being mustered into service for one year on June 30, 1861, to date from April 25. On July 25, they transferred to become Company K, 59th Virginia Infantry.

From these two groups, and a number of others from Kanawha and surrounding counties, the 1st Kanawha Regiment was organized on April 25, 1861. The exact composition of the regiment at that time is difficult to ascertain, but when brought to completion it apparently was constituted as follows. Many of these company designations would change when the 1st Kanawha became the 22nd Virginia Infantry,

probably in August of 1861, and again when the regiment reorganized on May 1, 1862.

1st Kanawha Regiment
May-August 1861

Border Rifles (Putnam County)
Charleston Sharpshooters (Kanawha County)
Kanawha Rangers (Kanawha County)—cavalry
Bruce Rifles (Greenbrier County)
Mountain Cove Guards (Fayette County)
Nicholas Blues (Nicholas County)
Elk River Tigers (Kanawha County)
Kanawha Riflemen (Kanawha County)
Boone Company (Boone County)
Kanawha Artillery (Kanawha County)—artillery

On April 29, 1861, Virginia authorities commissioned John McCausland a lieutenant colonel of state volunteers and placed him in charge of the situation in the Kanawha Valley. Having been raised and educated in the valley he had a familiarity with the land and its people. Additionally, he had a military background as an 1857 graduate (with first honors) of V.M.I., later serving as an assistant professor of mathematics at that institution. He organized the 1st Rockbridge Artillery and was elected captain, but turned down the command in order to return to the embroiled Kanawha Valley.

*The preceding is from The Virginia Regimental Histories Series, titled "22nd Virginia Infantry," by Terry D. Lowry.

Amidst the roar of 20th century Charleston this leafy glade abided on Kanawha Boulevard, which is the exact route of the famed James River and Kanawha Turnpike of Civil War times. The Kanawha Riflemen—later Co. H of the 22nd Virginia Volunteer Infantry—must have passed this spot many times bearing their muskets to war. In 2020, this monument was removed.

MEMORIAL FOR RIFLEMEN OF '61 IS DEDICATED

Far Removed From Scene of Battle Impressive Ceremonies Are Held.

CHILTON DELIVERS DEDICATORY ADDRESS

John B. McCausland Only Surviving Gray General, Among Notables Present.

(BY HERBERT PEAHLER)

Far removed from any scene of battle, with no suggestion of strife in the world, accompanied by no bugle notes nor a last salute, and militant only by the appearance of a few uniforms of gray, a memorial was dedicated yesterday afternoon at 4:30 o'clock at Kanawha park to the dead and surviving who composed the Kanawha Riflemen in the years of the rebellion.

In a sequestered spot in Kanawha park, just a few hundred feet from "some inland river," in the shadow of trees whose growth antedates the oldest living person in the valley, the Daughters of the Confederacy erected a monument and bronze tablet to the men of the Kanawha Riflemen, who, in a great crisis, when facing two alternatives, chose what to them was the proper course of conduct to be pursued.

Monument Is Unveiled

A perfect perspective was revealed yesterday when, amidst the pastoral setting, a few hundred men and women, their children and grandchildren, gathered beneath the kindly shade of the trees in the park and listened to some spoken words and then saw two children, direct descendants of the riflemen, pull some string which unveiled a bit of polished marble and a plate of bronze on which were engraved the names of the men who in 1856 made the great decision. And in the gathering were men who had worn the blue and men who had worn the gray, and descendants of both. Nothing could have so perfectly revealed the healing influence of time.

The distinguished guest on the occasion was General John B. McCausland, the last surviving general officer of the Confederate army and the man who recruited the Kanawha Riflemen. This 85-year-old veteran, who lives in a great mansion in the Kanawha valley in Mason county, came to Charleston to attend the exercises. While here he was the guest of former United States Senator William E. Chilton, who was the orator of the day. General McCausland also spoke, reviewing some incidents of a long and full life. He was with Stonewall Jackson at Virginia Military institute when both were instructors there, McCausland being a professor of mathematics. He was in many battles of the war, was with Lee when he surrendered, and was a globe trotter for many years. He recalled an incident when he witnessed Maximillian, who Napoleon tried to put on the Mexican throne, as he stood bareheaded in the palace in Mexico City a few months before he was backed up to a stone wall and shot by a firing squad.

But General McCausland was not the only living link that joined the old with the present. In the audience who listened to his address was the venerable John Q. Dickinson, 90 years of age, and the estimable Mrs. George S. Chilton, 89, both familiar with the history which was being perpetuated by the erection of the inanimate stone and bronze memorial.

General McCausland was retrospective, as becomes all aged men. He lived in the past; he revealed qualities of scholarship and a life so full that much has overflowed in the fullness of the years. He told some amusing experiences and was a bit biographical, and when he had finished was the recipient of many congratulations as the people surged forward to meet him.

A young boy, whose experiences during even the late war are confined to mere memories of the departure of troops and their return and the war activities at home, and a little girl, who has known no note harsher than that of the fireside, pulled the cords that unveiled the memorial. The boy was Fontaine Broun, Jr., son of Mr. and Mrs. Fontaine Broun and grandson of Major Thomas L. Broun, a charter member of the riflemen; and the girl was Margaret Ruffner Earwood, great-granddaughter of Captain David Lewis Ruffner and also a great-granddaughter of Captain Richard Q. Laidley.

Mr. Fontaine Broun received the memorial and introduced his son and little Miss Earwood, who did the actual unveiling. . . .

Reprinted from *The Charleston Gazette*, Sunday, June 4, 1922

BAHLMAN GIVES HISTORY OF MEN SERVING IN '61

Former First Sergeant of
"Fayetteville Rifles"
Writes Interesting
Account of Kanawha's
Part in Civil War

SAYS SOME NAMES ARE MISSING FROM MEMORIAL

William F. Bahlman of Kansas City and the first sergeant and the last Captain of Company K, 22nd Virginia Infantry (Fayetteville Rifles) who has been visiting Captain Abbitt here, has returned to his home. Before his departure he wrote the following short history of the Kanawha Riflemen in the Civil War. It will be of historical value and of particular interest to survivors of the "War Between the States."

In the account of the dedication of the Kanawha Riflemen memorial published in the Gazette of June 4, 1922, there occur several mistakes and omissions which in justice to the company, desire to correct.

The company was mustered in on May 8, 1861. In addition to the names on the memorial the following belonged

Sheffy Baldwin of Staunton, Virginia.

Andrew Mat Donnally.

Andrew Van Donnally.

Charles Quarrier. Died from a wound after the war.

Jeff Davis, colored, cook with Armstead.

Charlie Chewning was killed at Lewisburg by one of our own shells, Lieutenant Guy Carr was killed at Dry Creek, Aug. 26, '63.
May 23, 1862

Failed to Find Grave.

Lieutenant Alanson Arnold was killed in front of Richmond in June 1864. I found the lot in which he was buried in Hollywood Cemetery in Richmond, Va., but could not find his grave.

Where others were killed I do not know.

Charles Turner and Theodore Turner of Coalsmouth, now St. Albans died at Parisburg, Giles county, on the same day of typhoid fever.

William F. Bahlman of the Fayetteville Rifles—22nd Virginia Infantry.
COURTESY LARRY LEGGE, BARBOURSVILLE, WV

I do not know where other men of the company died. Probably some of them were missing and never accounted for.

The author of the article of June says there is no one of the company living to give the remainder of the war, but I am still living and know something about it.

General McCausland never took the 22nd Regiment to Fort Donelson, for it was never there. The 36th Regiment was there and lost heavily. The 22nd lost 140 men out of 394, or 371-2 per cent in the Lewisburg fight. I have this information from Noyes (Plus) Rand, adjutant.

Quartered at Lewisburg.

In the spring of '63 the regiment came out of winter quarters at Lewisburg and was detached to join General William E. Jones in an attempt to destroy the bridges on the Baltimore and Ohio railroad which was the main artery of communication between the west and the east. The attempt failed as our artillery could not hit the supports of the bridges.

The regiment marched 28 days out of 35 and part of the time went hungry. I saw men throw balls of dough into the fire and snatch them out before they were half baked. A number of the men made the march bare-footed.

The next fight was Dry Creek near White Sulphur Springs, which lasted one and one-half days. The first exchange of prisoners was made at Vicksburg, Miss. In this there were 65 of the 22nd. I know for I was put into command of the Virginians, nearly 400. We were ex-

changed on September 10, '62 the very day on which the 22nd was fighting at Fayetteville. The next regular fight was at Droop Mountain in Pocahontas county. Here the regiment lost 15 officers out of 24, or 61-2 per cent. Among them were Captain Laidley and Lieutenant Donaldson. Companies A and K had no officers left.

In '64 the brigade was sent east and put into Breckenridge's Virginia division. The 22nd was in the fight at Newmarket and Monocacy and got close to Washington.

Sent to Richmond

The brigade was then sent to Richmond and was in the fight at Tottapotamy Creek and the second Cold Harbor. Here a part of Hancock's force struck the 26th Virginia battalion and broke through. Finnegan was in reserve with his Florida brigade and made a counter charge driving Hancock's men back through the gap. As they went back they scared off Lieutenant Donaldson and 18 men of the Kanawha Riflemen. Some of these men were being taken to Elmira, N.Y. prison when the train had an accident and Johnnie Patrick and Creed Parks were killed. Albert Singleton had both leg broken and Joe Matthews' temple was badly hurt, leaving a scar that looked as if made by a hot gridiron. All four of these men were K. R.

' In the fall of '64 the division went through the campaign between Sheridan and Early in the Valley of Virginia and when the campaign was ended the division was about ended too. At Winchester, September 19, '64, the division lost 1,100 out of 2,500 or 44 per cent. Sheridan had more cavalry than Early had infantry. The regiment was also at Cedar Creek, October 19, '64 and I presume at Fisher's Hill also.

In the Spring of '65 what was left of the division was near Christiansburg and was disbanded there. Colonel Patton was mortally wounded at Winchester, a one-fourth pound piece of shell having passed through both hips.

Major Bailey was mortally wounded at Droop Mountain and Lieutenant Colonel Barber was wounded at Dry Creek.

I could tell many incidents of the regiment but refrain. I have written this article entirely from memory as I have no diaries, data or memoranda, but I have no prejudices [stet]. There are a few survivors of the regiment and all of them are old and some of them feeble.

Reprinted from *The Charleston Gazette*, Sept. 17, 1922.

Company Orders #1 for the Kanawha Riflemen

John Rundle, a Kanawha Rifleman, published the *Kanawha Valley Star*, an early Charleston newspaper. It was first published in Buffalo as the *Star of the Kanawha Valley* in 1855. The paper moved to Charleston in 1857, where it took a strong Democratic and pro-Southern stand. Union troops confiscated Rundle's paper in 1861. TERRY LOWRY

Henry D. McFarland, member of the Kanawha Riflemen. His brother was president of the Bank of Virginia in Charleston. TERRY LOWRY

KANAWHA VALLEY STAR
April 30, 1861

Kanawha Riflemen

Company Orders #1
April 26, 1861

1. In compliance with the requisition of a Proclamation of the Governor of Virginia dated at Richmond the 19th of April 1861, this command will hold itself in readiness for marching orders.

2. In case such orders shall arrive, each one must provide himself with the following articles at least in addition to dress and fatique uniforms, to wit: two shirts, four collars, two pair of socks, two pair of drawers, one blacking brush and box (to any two files), two pair white Berlin gloves, one quart tin cup, one white cotton haversack, one case knife, fork and spoon, two towels, two hankerchiefs, comb and brush, and toothbrush. Some stout linen thread, a few buttons, paper of pins and a thimble, in a small buckskin or cloth bag.

3. There being no knapsacks in the possession of the company one ordinary sized carpetsack will be allowed to every two men, for the purpose of holding such of the above articles as are not in constant use. The knife, fork, spoon, haversack and tin cup, must be worn about the person, the first three and the last articles to the waist belt. Immediately after the receipt and promulgation of marching orders, the carpetsacks, duly packed, must be delivered to the Quartermaster Sergeant, neatly marked with the names of the two owners. Each file will procure a comfortable blanket and upon the receipt of orders, send the same in to the Quartermaster Sergeant, shaped into a neat and compact bundle conspicuously marked with his name.

4. It is earnestly recommended that all under clothes should be woolen, especially the socks, as cotton socks are utterly unfit for marching in, and all files should wear woolen undershirts. Shoes, sewed soles, and fitting easily, but not too loosely to the foot, coming up over the ankle, are infinitely preferable to boots, and should be made strong and servicable.

5. By the liberality and patriotism of the residents of Charleston (one of them a lady) flannel cloth (grey) has been furnished for fatigue Jackets, and provision made for cutting them, all members of the company are hereby required at once to have their measures taken and Jackets cut by Mr. James B. Noyes, tailor. Many ladies have kindly undertaken to make them up. All members of the company are required to have their Jackets finished by Wednesday afternoon next at the latest. By like liberality of another resident, cloth for haversacks has been procured, and they have been cut out by another lady. They will be delivered by the Quartermaster Sergeant, and they must be finished by the same evening.

6. Assistant Surgeon Joseph Watkins will immediately, upon the receipt of marching orders, prepare and put in portable form an ample supply of medicines, and be prepared to hire medical aid whenever required on the march or in transit or in camp. He will also provide himself with appropriate instruments, & c. In this connection the undersigned gratefully acknowledges on behalf of the company the liberal offer of a citizen of this town, to furnish free of charge all medicines required.

7. Quartermaster Sergeant John Dryden will immediately, on the receipt of marching orders, procure the necessary transportation for the baggage of the command and necessary camp utensils and fixtures, and in case the order shall require a march overland, will lay in at least six hundred rations, and provide for their transportation; the ration being one pound and a half of pork or beef (as much of the latter as can be purchased fresh), eighteen ounces of meat and one-fourth pound of corn meal, and to each one hundred rations the following articles, ten pounds of rice, six pounds of coffee, twelve pounds of sugar, one gallon of vinegar, one pound of star candles, four pounds of soap and two quarts of salt. Private Joseph M. Brown is hereby detailed as an assistant to the Quartermaster and will report to him accordingly.

8. The band will go as a band, and are as such until further orders, will carry their instruments with them, but in every other respect will govern themselves by the preceeding directions.

9. The undersigned, in issuing these preparatory orders has but little doubt that the services of this command will be required to aid in driving the invader from the soil of Virginia, but has none that every Rifleman will respond cheerfully and with alacrity to the call of his State, and be prepared to do his duty bravely under the grand old flag of Virginia.

GEO. S. PATTON
Captain

KANAWHA RIFLEMEN

In Charleston on Kanawha Boulevard is a small public park, once the private cemetery of the Ruffner family. It contains a memorial to the Kanawha Riflemen, which bears the following roster of names:

THIS MEMORIAL
ERECTED
BY THE KANAWHA RIFLEMEN CHAPTER
UNITED DAUGHTERS OF THE CONFEDERACY
IN HONOR OF THE KANAWHA RIFLEMEN
FIRST ORGANIZATION OF THE COMPANY 1856

Captain--George S. Patton
First Lieutenant-----------------------------------Andrew Moore
Second Lieutenant-------------------------Nicholas Fitzhugh
Third Lieutenant-----------------------Henry D. Ruffner

SECOND ORGANIZATION 1858

Captain--David L. Ruffner
First Lieutenant-----------------------------------Richard Q. Laidley
Second Lieutenant-----------------------------------Gay Carr
Third Lieutenant--------------------John P. Donaldson

THIRD ORGANIZATION 1861

Captain--Richard Q. Laidley
First Lieutenant-----------------------------------John P. Donaldson
Second Lieutenant-------------------------Henry W. Rand
Third Lieutenant-----------------------Alanson Arnold

NON-COMMISSIONED OFFICERS AND PRIVATES

Arnold, E. S.	Chewning, Charles	Lewis, Joel S.
Barton, Norman	Clarkson, A. Q.	Lewis, John
Blaine, Charles	Cook, Walton	McQueen, Archibald
Boswell, Martin	Cox, Frank	McFarland, Henry D.
Brodt, J. T.	Cushman, William	McMullen, John
Bradford, Henry	Doddridge, J. E., Jr.	McClelland, Robert
Brooks, W. B.	Doddridge, Philip	Malone, William
Broun, Thos. L.	DeGruyter, M. F.	Mathews, John
Broun, Jo. M.	Fry, James H., Jr.	Miller, Samuel A.
Brown, Siline	Grant, Thos. T.	Miller, H.
Cabell, H. Clay	Hale, John P.	Noyes, Benjamin
Caldwell, William	Hansford, Carroll M.	Noyes, Frank
Carr, Gay	Hare, Robert	Noyes, James B.
Carr, John O.	Hopkins, --------	Noyes, James B., Jr.
Chambers, John	Lewis, James F.	Noyes, William
Noyes, John	Ruby, John C.	Summers, William S.
Parks, Cecil	Rundle, John	Summers, Geo. W., Jr.
Parks, Bushrod	Ruffner, David L.	Swann, John S.
Patrick, A. S. Dr.	Ruffner, Daniel, Jr.	Swann, Thomas B.
Patrick, John	Ruffner, Joel, Jr.	Teays, Stephen T.
Quarrier, Joel S.	Ruffner, Meridith P.	Thompson, Cameron L.
Quarrier, William A.	Ruffner, Andrew L.	Thompson, Thornton
Quarrier, Monroe	Shrewsbury, Andrew	Turner, Benjamin F.
Rand, Noyes	Shrewsbury, Joel	Watkins, Joseph F.
Read, Fred M.	Spessard, Jacob	Wehrle, Meinhart
Reynolds, Fenton M.	Smith, Isaac Noyes	Welch, George L.
Reynolds, William	Smith, Thomas	Welch, Levi
Roberts, Thomas	Singleton, Albert	Welch, James
Ruby, Edward	Snyder, W. B.	Wilson, Henry
	Smithers, David	Wilson, W. A.

Dedicated to those who served in the Confederate Army—1861-1865

Embossed on the memorial plaque is the name of William Armistead, colored cook, faithful during the war. "Uncle Billy" was a cook with the Kanawha Riflemen, who served throughout the war. He actually fought with the 22nd Virginia Infantry at the Battle of New Market in 1864. After the war, he came back to Charleston and lived there for many years.

This photo appeared in an April 17, 1938, issue of *The Charleston Daily Mail*. It shows a group of just-returned Kanawha Riflemen officers in 1865. As the article's author, George W. Summers, stated: "Despite their beards, evidently grown when shaving facilities were unknown in camp, these fellows were scarcely more than boys when the picture was made. For many years after the end of the Civil War some of these 'boys' continued to live here, though none of them is living now." Standing, left to right: Maj. Thomas L. Brown, Col. Thomas Smith, Maj. Samuel A. Miller, Col. William Fife [Summers incorrectly named him Thomas E. Fife] and Capt. Nicholas Fitzhugh. Seated, left to right: Dr. Joseph Watkins, Lt. William A. Quarrier, Capt. Richard Q. Laidley and Capt. John Swann. COURTESY WEST VIRGINIA STATE ARCHIVES

Kanawha Rifleman Samuel A. Miller had a varied career before and after the war. He was born in Mt. Jackson, Va., in 1820. He graduated from Gettysburg College and after extensive travels to Europe and Mexico moved to Mason County, Va. (now W. Va.). In 1840 he moved to Charleston and studied law with George W. Summers, entering into practice with him in 1844. He became associated with the Ruffner, Donnally & Co. salt firm in 1851 and was president of the company until it went out of business. Enlisted in the Riflemen in 1861 and appointed A. Q. M. in August 1861. Then A. D. C. on McCausland's 2nd brigade in May 1862 and then appointed Brig. Q. M. of Patton's 2nd brigade in November 1862. In February 1863 he resigned from the army and was elected to the Confederate Congress, filling the unexpired term of Gen. A. G. Jenkins. He was reelected in 1864 and served till the end of the war. After the war he fled to Canada briefly, but soon returned to Charleston. After receiving a pardon from President Johnson and regaining his law practice, he went into partnership with his brother-in-law, William A. Quarrier. He was in the House of Delegates in 1874–75 and in a law partnership with ex-governor E. W. Wilson. Miller died in 1890 and is buried in Spring Hill Cemetery.

Charleston's Boys in Gray

William F. Fife graduated from V.M.I. in 1856. As a lieutenant colonel of the 36th Virginia Infantry he was wounded at the Battle of Cedar Creek in 1864. Fife was killed in a train wreck below the present site of Dunbar on July 4, 1891, and was buried in Spring Hill Cemetery.

Thomas Smith was the son of William "Extra Billy" Smith, a former governor of Virginia. He served in the 22nd and later as colonel of the 36th Virginia Infantry. He was wounded at the Battle of Cloyd's Mountain. After the war, he became chief justice of the New Mexico Territory. He died in Warrenton, Va., in 1918.

John S. Swann was born in 1822. He became a lawyer and was practicing law in Charleston in 1860. He enlisted in the Kanawha Riflemen in 1861 and later transferred to the 59th Virginia. He served most of the war with Edgar's 26th Battalion Virginia Infantry. Swann died in 1903 and is buried in Spring Hill Cemetery.

Joseph F. Watkins graduated from the University of Virginia in 1856 and from Jefferson Medical College in 1858. He practiced medicine in Charleston until enlisting in the Kanawha Riflemen in 1861. He transferred to the 36th Virginia as a doctor. After the war, Watkins practiced medicine in Charleston until his death in 1888. He is buried in Spring Hill Cemetery.

Nicholas Fitzhugh was born in 1823. He attended Mt. Ovis Academy in Charleston, where prior to the war he practiced law. Fitzhugh was a second lieutenant in the 22nd and then went on to become a major and adjutant in Jenkins' Cavalry. He mustered out at Appomattox. He is buried in Spring Hill Cemetery.

William A. Quarrier was born in 1828. He attended Mercer Academy in Charleston, where he practiced law prior to enlisting in Hale's Artillery in 1861. He served in the Commissary Department at Saltville, Va., until the end of the war. After the war, he again practiced law in Charleston and also ran for the U.S. Senate. He died in 1888 and is buried in Spring Hill Cemetery.

Biographical information from Terry Lowry's "22nd Virginia Infantry" and John Scott's "36th Virginia Infantry"

George Smith Patton
1833-1864

George Smith Patton, born in Virginia in 1833, was the greatgrandson of Gen. Hugh Mercer, who died a hero in the American Revolution at the Battle of Princeton.

Mercer County, W.V., and its county seat of Princeton commemorate the general as did the renowned Mercer Academy in Charleston. He also received a land grant that included the present site of Sissonville for his service in the French and Indian War.

Young George S. Patton was raised in the military tradition graduating from the Virginia Military Institute in 1852.

For a few years Patton taught mathematics and English at a small Richmond academy. In 1856 he, along with his new bride, Sue Glassell, moved to Charleston, Virginia, where he formed a law partnership with Thomas Broun.

As war clouds gathered Patton's V.M.I. training was put to good use when in 1856 he formed a company of Charleston aristocrats into a military organization called the Kanawha Riflemen.

When Virginia seceded in 1861 Federal troops from Ohio invaded the Kanawha Valley. They were met by Patton and the Kanawha Riflemen, who fought alongside other Confederate forces at the Battle of Scary Creek on July 17, 1861. Patton was severely wounded in his right arm as the Confederates retreated from Charleston. He was left at his home "Elm Grove" near Quarrier and Dunbar streets to recuperate. After a few weeks of looking out his windows at the passing Federal troops, Patton was paroled and allowed to go to his old family plantation home near Richmond. The wound healed poorly but after nearly a year his "exchange" came through and he returned to the Confederate Army.

In May 1862 Colonel Patton and the 22nd Virginia, of which Company H was the former Kanawha Riflemen, took part in the campaign against Federal forces, who were attempting to cut rail lines in southwest Virginia.

At the Battle of Giles Court House Patton was again wounded and contracted blood poisoning, which proved worse than the actual wound. He returned to Richmond until September, when as colonel of the 22nd he joined other Confederate forces ordered by General Lee to retake the Great Kanawha Valley with its precious salt works.

In a memorable three days, September 10 to 13, 1862, the Confederate Army under Gen. W. W. Loring drove the Federals out of the valley in headlong retreat.

After less than a month the hometown Charleston boys of the 22nd were forced to withdraw to Lewisburg, where they stayed all winter.

In the summer of 1863 Col. Patton and his men won a splendid victory at Dry Creek near White Sulphur Springs against Union Gen. Averell. At last Patton could take his place as a successful commander alongside his illustrious ancestors. Sadly at the same time, news came that Patton's brother Tazewell had died at Gettysburg in Pickett's Charge.

In November Patton met defeat by the overwhelming numbers of Federals in the Battle of Droop Mountain.

In May 1864 Patton and his 22nd fought in the Battle of New Market alongside the fabled boy soldiers of V.M.I., who fought gallantly. Patton's horse was shot from under him, but the battle was a clear Confederate victory.

Like his greatgrandfather in 1777 Patton was not to survive the war. After a last gallant strike at the Federals, which saw the 22nd Virginia camped within sight of Washington, D.C., under Gen. Jubal Early's command, Patton met his fate on Sept. 19, 1864, in the Third Battle of Winchester. A piece of shrapnel struck his leg as he sat astride his horse.

Just as he had done with his arm in July 1861, Patton refused to allow his leg to be amputated. It appeared he would recover, but gangrene set in after a few days.

On Sept. 25, 1864, the dark star that seemed to raise the Patton men to glory but condemn them to death ended the life of 32-year-old George Smith Patton, late of Richmond by way of Charleston.

Col. Patton fathered several children, among whom George William, 1856–1927, became the father of one of the most illustrious figures in 20th century military history—Gen. George Smith Patton Jr. commander of the U.S. Third Army.

From Hugh Mercer of the Revolutionary War to George S. Patton of the Civil War to George S. Patton Jr. of World War II, the history of the world was changed by these brave men.

The defeat of the Nazi army was thus born in the cradle of old Virginia.

George Smith Patton (1833–1864), a pre-war view.

Sue Glassell Patton, a post-war view.

Dear Brother . . . I send you a brief . . . account of The Action of Scarey Creek.

(Patton's letter to his brother is on deposit at the Huntington Library, San Marino, Calif. Patton referred to himself in the third person)

Putnam Co. Va. July 17, 1861

. . . at the mouth of the Coal 12 miles below Charleston under Major Geo. S. Patton. Major Patton was then on the opposite side of the Kanawha River, & 10 to 12 miles below the nearest Confederate forces. The enemy had moved a column from Guyandotte which compelled Major Patton to send a large portion of his force down that road.

As some of the Federals advanced in force Major Patton burnt the bridge & placed a picket there. Subsequent reflection satisfied him that it was a good place for a stand as both flanks of his small force would be, in a measure, protected by the river & hills—and he gradually threw nearly his whole disposable force there. . . the [Union] army had in fact crossed the river but it was only a *ruse*, and about noon the same day the Federal commander threw over the 12th Ohio Regiment Col. Lowe, a large portion of the 21st Ohio Regiment Col. Norton, a section of Artillery (two 10 lb. rifled pieces) and a company of cavalry, in all about 1,500 men; with orders to march upon & route the small force at Scarey.

The action was commenced shortly after two o'clock— the first guns being fired by the Confederate Artillery . . . The action soon became general—and in a few moments the two cavalry companies arrived, were dismounted, and Capt. Lewis thrown into the woods to the left to prevent a [sic] annoyance by a flanking party, while Capt. A. C. Jenkins' . . . men were held in reserve.

. . . our men were gradually being forced back and fell into some confusion—Sweeney still held his houses, but the odds against him forced him to fall back— . . . with a shout our men charged—drove them back across the creek—beyond the houses— & back to their original position. In the struggle Major Patton was severely wounded in the shoulder and was forced to retire a short distance to the rear.

The Confederate loss was 3 killed and 9 wounded of which two died of their wounds. The yankees left 12 or 15 dead on the field, but by their own confession their loss was not less than 200 killed and wounded.

Late in the evening Col. Woodruff of the 2nd Kentucky— Col de Villiers of the 11th Ohio—Lt Col Neff—two Captains of the 2nd Kentucky, who strong in their faith of Yankee invincibility, and knowing our weak numbers—had ridden up to see the "rebels crushed," were captured, & spent many months in the "Libby."

The affair is chiefly remarkable as being fought so early in the war, against such odds of numbers and arms (for be it recollected we never had over 400 actually engaged, & they chiefly with mountain rifles & "flintlocks") and almost in sight—certainly in full hearing of Cox's whole army. These mountain men with—in many instances cartridges in their pockets, just organized & underdrilled—whipped 4 times their number of armed and disciplined Yankees & put them to a shameful and disgraceful fight—In the open field they met them face to face and conquered.

Upper left: A wartime photo of Patton with beard. Many photos of Patton recently published show his profile reversed. This is the correct view. Lower left: Patton's grave stone at Mt. Hebron Cemetery in Winchester, Va. Upper right: Gen. George S. Patton Jr., grandson of Confederate Colonel Patton, won fame in World War II battles of North Africa, Sicily and continental Europe. PHC

A drawing of the Kanawha Riflemen's uniform, designed by George S. Patton.

The Craik-Patton Home

Elm Grove was located on the site of the old Kanawha Valley Hospital until 1906 when it was moved to 1306 Lee Street. In 1971 it was moved to Daniel Boone Park and restored to its 1834 appearance by the National Society of Colonial Dames of America in West Virginia. James Craik, grandson of George Washington's personal physician, built the house. George S. Patton and his wife, Susan, bought the house in 1858 shortly after their son George S. Patton Jr. was born. After the senior Patton joined the Confederate Army his wife moved to Virginia. She sold the house to Andrew Hogue, and it was in this family until 1921.

A postwar color tinted postcard of a wartime view of the 8th West Virginia Infantry (later the 7th West Virginia Cavalry) at Progress Mills at Buffalo, West Virginia. The card is titled 7th West Virginia Regiment, leading many to wrongfully identify this as the 7th West Virginia Infantry, a regiment that was never in the area during the war. The view was taken in Buffalo, Putnam County, possibly in 1862. It is one of the few wartime photographs to be discovered from the Kanawha Valley area. A blow-up of the photo shows the regiments band and a better view of its members' uniforms. Progress Mills on the right once stood on the front street in Buffalo. Soldiers were quartered in the building, which had standing water in the basement. Three of the soldiers died from typhoid fever, carried by mosquitoes that bred in the basement water. PHC

Reminiscences of Maj. Thomas L. Broun

Thomas L. Broun, son of Edwin Conway Broun, was born in Loudoun County, Va., Dec. 25, 1823. He graduated at the University of Virginia in 1848, and taught school for two years. He studied law with Hon. G. W. Summers in Charleston, Kanawha County, and was admitted to the bar in 1852. He succeeded William S. Rosecrans as President of the Coal River Navigation Company in 1858. His law partner was George S. Patton.

Maj. Broun entered the Confederate service as a private in April 1861, and continued in service until the surrender at Appomattox, becoming major in the Third Regiment of the Wise Legion. During the campaign on Big Sewell Mountain, in western Virginia, in the fall of 1861, he was very ill from typhoid pneumonia, which rendered him unfit for field service. Consequently he was stationed at Dublin Depot, Va., as commandant of the post and major quartermaster. Maj. Broun was very dangerously wounded May 9, 1864, in battle at Cloyd's Mountain. Gen. A. G. Jenkins, Maj. Thomas Smith, and Maj. Thomas L. Broun were all wounded and taken to the residence of Mr. Guthrie, residing near Dublin Depot. The evening after the battle a squad of Federal cavalry went to Mr. Guthrie's house and paroled Gen. Jenkins and Maj. Smith. The Federal surgeon said to his clerk, in the presence of Maj. Broun, "Report him dead, for he will die tonight"; and so Maj. Broun was reported as killed in that battle. [Broun survived his wounds and after some confusion as to his identity—for he had been reported dead—he was paroled.]

Early in June, 1865, Maj. Broun returned to Charleston (after an absence of four years), and found that his valuable law library and other personal property had been confiscated. There were, besides, several indictments in the United States Court against him as a recruiting officer in Boone and Logan counties for the Confederate army under orders from Gen. H. A. Wise, of the Wise Legion, and Gov. Letcher, of Richmond, Va. Officials of the new State of West Virginia were bitterly opposed to the so-called Rebels returning to their homes, and Maj. Broun was ordered to appear before Gen. Oley, then in command of troops stationed here. He presented his parole from Gen. Carroll, and claimed protection, but this was not granted him. Thereupon the friends of Maj. Broun and other Confederates telegraphed to Gen. Grant, who at once replied that the paroled soldiers and officers of the Confederate army should have all rights and privileges granted to them by their paroles, without any interference whatever by State or local United States authorities. This reply brought much quiet and satisfaction to Confederates in West Virginia.

Maj. Broun was married in Richmond, June 8, 1866, to Miss Mary M. Fontaine, daughter of Col. Edmond Fontaine of Beaver Dam, Hanover County, Va.

As ex-Confederate soldiers were not permitted to practice law in West Virginia for some time after the war, Maj. Broun moved to New York City in June, 1866, where he resided for several years. After political disabilities of West Virginia were removed, he returned and resumed the practice of his profession in Circuit Courts, in the Supreme Court of Appeals of West Virginia, and in the United States District and Circuit Courts for West Virginia. As a delegate from the Diocese of West Virginia he attended the Triennial Episcopal Conventions held in New York, Philadelphia, and Chicago, in 1880, 1882, and 1886.

Taken from *Confederate Veteran* magazine, 1901.

Thomas Lee Broun was a member of the Kanawha Riflemen, an 1848 graduate of the University of Virginia, a school teacher and a law partner of George S. Patton. He succeeded William S. Rosecrans as president of the Coal River Navigation Company in 1858. TERRY LOWRY

On a high ridge overlooking the city where he lived is the grave of Thomas L. Broun. Many of the old soldiers of the Confederacy rest in Spring Hill Cemetery, including various regiments—the "Kanawha Riflemen," 22nd Virginia Volunteer Infantry and 36th Virginia Volunteer Infantry among others. Spring Hill Cemetery of Charleston is on the National Register of Historic Places and can be reached from the north end of Morris Street.

General Lee acquires "Traveller"

(This is the story of General Lee's acquisition of his famous horse, "Traveller," as written by Thomas L. Broun of Charleston to *The Greenbrier Independent*, on Aug. 6, 1900, and repeated by Shirley Donnelly in the *Beckley Post-Herald*, Nov. 24, 1962.)

"He was raised by Andrew D. Johnston near the Blue Sulphur spring of 1861. He was of the 'Gray Eagle' stock, and, as a colt, took the first premium under the name of Jeff Davis at the Lewisburg fairs for each of the years 1859 and 1860.

"He was four years old in the spring of 1861. When the 'Wise Legion' was encamped on Sewell Mountain opposing the advance of the Federal army under General Rosecrans in the fall of 1861, I was a major to the Third Regiment of Infantry in that Legion, and my brother, Capt. Joseph M. Broun, was quartermaster to the same regiment. I authorized my brother to purchase a good serviceable horse of the best Greenbrier stock for our use during the war.

"AFTER MUCH INQUIRY and search he came across the horse above mentioned and I purchased it for $175—gold value—in the fall of 1861, of Capt. James W. Johnston, son of the A. D. Johnston first above mentioned.

"When the Wise Legion was encamped about Meadow Bluff and Big Sewell mountains I rode this horse which was then greatly admired in camp for its rapid, springy walk, high spirit, bold carriage and muscular strength. He needed neither whip nor spur and would walk his five or six miles an hour over the rough mountain roads of West Virginia, with his rider sitting firmly in the saddle and holding him in check by a tight rein, such vim and eagerness did he manifest to go right ahead so soon as he was mounted.

"WHEN GENERAL LEE took command of the Wise Legion and Floyd Brigade that were encamped at and near Big Sewell Mountains in the fall of 1861, he first saw this horse and took a great fancy to it. He called it his colt and said he would need it before the war was over. Whenever the general saw my brother on this horse he had something pleasant to say to him about 'my colt,' as he designated this horse.

"As the winter approached, the climate in the West Virginia mountains caused Rosecrans' army to abandon its position on Big Sewell and retreat westward. General Lee was thereupon ordered to South Carolina. The Third Regiment of the Wise Legion was subsequently detached from the army in Western Virginia and ordered to the South Carolina coast where it was known as the sixtieth Virginia Regiment under Colonel Starke.

A post-war view of Lee and Traveller.
WASHINGTON & LEE UNIVERSITY ARCHIVES

"UPON SEEING my brother on this horse near Pocotaligo, in South Carolina, General Lee at once recognized the horse and inquired of him pleasantly about his colt. My brother then offered him the horse as a gift, which the General promptly declined and at the same time remarked:

" 'If you will willingly sell me the horse I will gladly try it for a week or so to learn its qualities,' whereupon my brother had the horse sent to Gen. Lee's stable. In about a month the horse was returned to my brother, with a note from Gen. Lee stating that the animal suited him but that he could no longer use so valuable a horse in such time unless it were his own; that if my brother would not sell, please keep the horse; with many thanks. This was in February, 1862.

"AT THAT TIME I was in Virginia on the sick list from a long and severe attack of camp fever contracted in the campaign on Big Sewell Mountains. My brother wrote me of Gen. Lee's desire to have the horse and asked me what to do. I replied at once: 'If he will not accept it, then sell it to him at what it cost me.' He then sold the horse to General Lee for $200 in currency, the sum of $25 having been added by General Lee to the price I gave for the horse in September, 1861, to make up for the depreciation in our currency from September, 1861, to February, 1862.

"In 1868 General Lee wrote to my brother stating that this horse had survived the war—was known as Traveller—and asking for its pedigree, which was obtained, as above mentioned, and sent by my brother to General Lee."

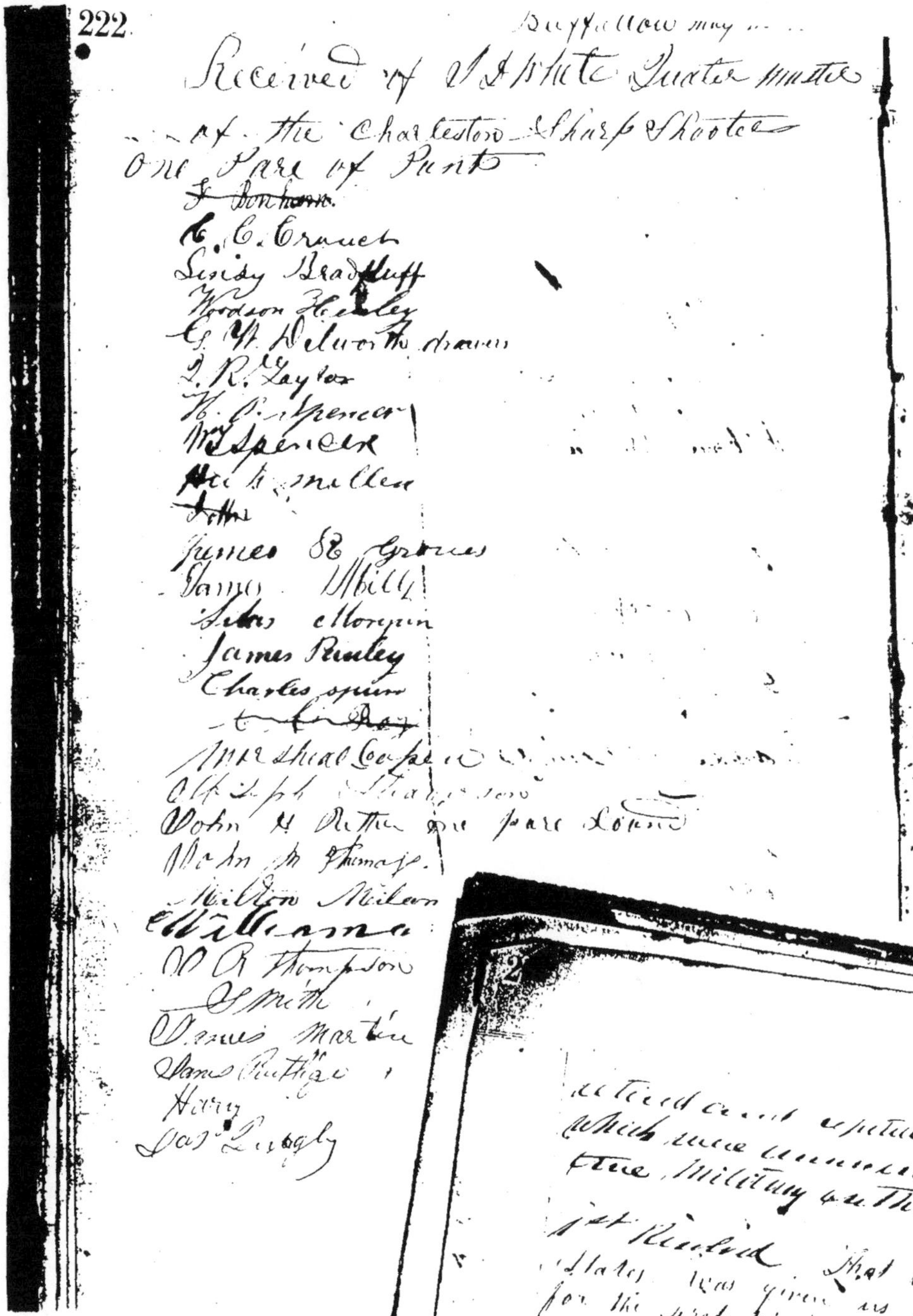

222

Buffallow may ...

Received of J A White Quarter master of the Charleston Sharp Shooters One Pare of Pants

C. C. Crauch
Linsey Bradfluff
Woodson Hensley
G. W. Delworth drawn
L. R. Taylor
H. C. Spencer
Wm Spencer
James E. Groves
James Mills
Silas Morgan
James Rinley
Charles Spurr
Marshal Cooper
John H. Rutter one pare found
John M Thomas
Wilson Milam
Williams
W R Thompson
Smith
James Martin
James Rutlidge
Harry
Jas Lusgly

Roll book of the Charleston Sharpshooters showing the quartermaster distribution of pants to its members at Buffalo on May 15, 1861.
WVSA

Declaration from the Charleston Sharpshooters roll book:

1st Resolved. That the union of these states was given us by our great forefathers for the protection of our liberties our persons and our property from foreign or domestic violence and whilst it continues to give that protection it shall have our firm support but when it comes a union wherein rather the liberty the persons or the property of the people of this mother countrywealth revolves upon her own sons alone and her sister states of the south for protection the union is already at an end and we should look down upon it as the brave mountaineer looks down upon his foes in the plain preparing to scale the rugged heights by his freedom and repose.

received and adopted the following resolutions which were unanimously adopted with true military enthusiasm

1st Resolved That the union of these States was given us by our great forefathers for the protection of our liberties our persons and our property from foreign or domestic violence and whilst it continues to give that protection it shall have our firm support but when it comes a union wherein rather the liberty the persons or the property of the people of this mother commonwealth revolves upon her own sons alone and her Sister States of the South for protection the union is already at an end. We should look down upon it as the brave mountaineer looks down upon his foes in the plain preparing to scale the rugged heights of his freedom and repose

2nd Resolved That we trust the true men of the north may rally to the ...

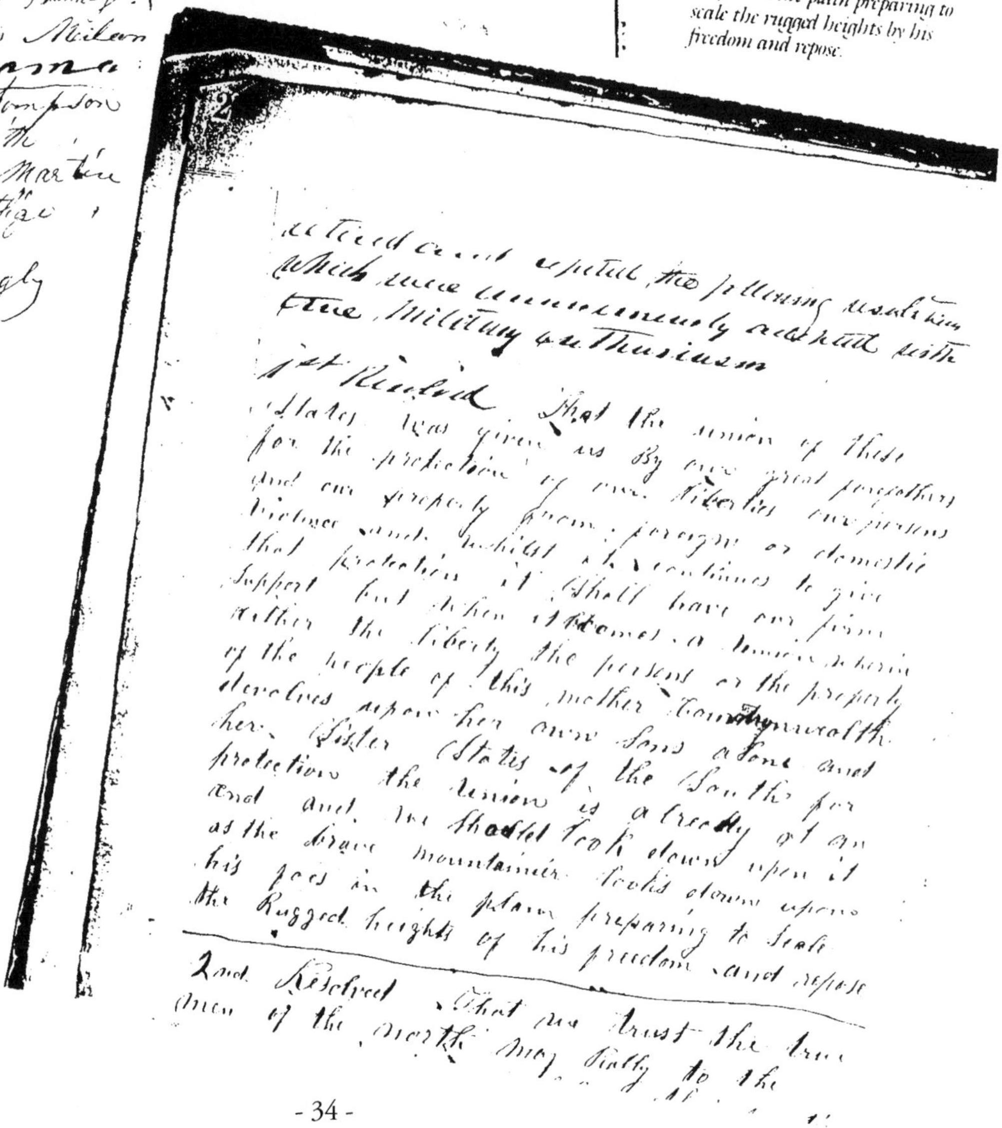

Kanawha County's Confederate General Birkett Davenport Fry was born in Coalsmouth in 1822. His mother was one of the earliest school teachers. She was a sister of Col. Philip Roots Thompson, one of the founders of Coalsmouth. He graduated from VMI, but failed out of West Point in 1844. Fry served in the Mexican War, later moved to California and participated in an expedition to Nicaragua in 1855. He then moved to Alabama and managed a cotton mill until the start of the war. Fry became a colonel of the 13th Alabama Infantry and was appointed an aide to Stonewall Jackson   May 1861. He was wounded in action at Seven Pines, Chancellorsville and Sharpsburg. The general commanded Archer's Brigade at Gettysburg where he was captured on July 3. He was exchanged in April 1864 and appointed brigadier general in May 1864. He faced General Sherman at Augusta, Ga. At the end of the war he spent three years in exile in Cuba and then returned to Alabama and was elected superintendent of schools. Later he returned to the cotton mill business in Alabama and Virginia. He died in 1891.
USAMHI

Other Units of Virginia Volunteers

The Kanawha Artillery was possibly organized in the late 1850s and called in state service in 1861. Dr. John P. Hale largely organized and financed the artillery with William A. Quarrier and James Welch as lieutenants. The artillery was at the surrender of Fort Donaldson but a number refused and escaped. Some members transferred to Capt. Thomas E. Jackson's Virginia Horse Artillery.

The Charleston Sharpshooters was organized in December 1859 and mustered into state service on June 30, 1861, and into Confederate service on Aug. 1, 1861, as Co. K, 59th Regt. Va. Inf. Assigned as Co. A, 26th Bn. Va. Inf. on June 17, 1862. Capt. John S. Swann was in command.

Capt. (Dr.) Andrew R. Barbee

The Elk River Tigers enlisted on June 8, 1861, and reorganized on May 1, 1862. Led by Capt. Thomas B. Swann and later George S. Chilton.

The Putnam County Border Riflemen (Rifles) was led by Capt. (Dr.) Andrew Russell Barbee and fought at Scary Creek.

The Kanawha Rangers was Co. K, 22 Regt. Va. Vol. under Captain Thomas J. Huddleston. Charles I. Lewis was elected captain of the company to fill the vacancy caused by the death of Captain Huddleston. The original company was enlisted on May 27, 1861, by Col. C. Q. Tompkins.

The Coal River Rifle Company organized in December 1859 with men from Kanawha and Putnam counties. Little is known of this unit and could have evolved into the Kanawha Militia.

Col. Thomas B. Swann

Lt. Col. John J. Polsley
and his 7th West Virginia Cavalry

The 8th Regiment Virginia Volunteer Infantry was organized in the fall of 1861 at Charleston, West Virginia. John J. Polsley became Adjutant of the regiment in December of the same year. In April 1861 it became a part of General Frémont's Mountain Department and with the 60th Ohio Infantry formed the advance brigade in Frémont's pursuit of Stonewall Jackson up the Shenandoah Valley. In the same year the regiment served admirably under Gen. Pope during his disastrous campaign that ended with the Union general being hurled back against the defenses of Washington. Returning to the Kanawha Valley, the 8th Regiment became a part of the Fourth Separate Brigade in March 1863, and for the third time was involved in an inglorious retreat by being forced across the state to Clarksburg. It was not until the regiment came under the command of Gen. William Averell in May 1863, that it met with aggressive and strategic leadership. Under Averell, the regiment—now known as the 8th West Virginia Mounted Infantry—took part in two major campaigns, the first into Maryland where an attempt was made to intercept Lee on his retreat from Gettysburg. The second campaign, ending in December 1863, was in the nature of a raid on Confederate installations in southern Virginia.

Early in 1864 the 8th Regiment was reorganized at Charleston and was converted into the 7th Regiment West Virginia Cavalry. From April to July 1864 the unit served under Maj. Gen. George Crook and took part in several raids on enemy fortifications in Virginia. From July 1864 until being mustered out in August 1865 the regiment served on patrol duty in West Virginia. Polsley was captured in December 1863 and confined in Confederate prisons until he was exchanged in September 1864. Returning to his regiment, he served with that unit until mustered out in August 1865 as a lieutenant colonel. He died suddenly at his home in Charleston in December 1866 and is buried in Charleston's Spring Hill Cemetery.

Grave marker for Lt. Col. John J. Polsley of the 7th West Virginia Cavalry. The preceding is a brief history of Polsley and the 7th as presented for a Master of Arts degree thesis by Eugene Wise Jones in 1949 at the University of Akron.

Col. John H. Oley.

Officers of the 7th West Virginia Cavalry. Col. John Hunt Oley is in the center of the photograph. WVSA

John Hunt Oley was born in New York in 1830. He was a member of the 7th Regiment of the New York National Guard. In the fall of 1861 he organized the 8th Regiment, Virginia Infantry in the Kanawha Valley, with headquarters in Charleston. It was mustered into Federal service on Oct. 29, 1861, with Oley as a major. Later the regiment became the 7th West Virginia Cavalry. Oley was promoted to lieutenant colonel in October 1862 and colonel in March 1863. The regiment was stationed in the Kanawha Valley in 1864–65 and was mustered out of service in August 1861. Colonel Oley was made a brevet brigadier general in March 1865. He was appointed Commissioner of Internal Revenue in 1871. He moved to Huntington upon its founding and served as city recorder and a member of the school board. He was a prominent citizen of the new city until his death in March 1888. He is buried in Huntington's Spring Hill Cemetery. It is merely a coincidence that both Huntington and Charleston have cemeteries named Spring Hill.

Colonel Oley's grave.

Shoulder boards of Lt. John McCombs, of the 7th West Virginia Cavalry. These have been kept in a Putnam County farmhouse since the war.

The Buffalo Guards

Company A, Thirty-Sixth Virginia Confederate Infantry, was enlisted at Buffalo, Putnam County. It entered service in May 1861, and disbanded at Christianburg, Virginia, April 12, 1865, three days after the surrender at Appomattox. Among its casualties were: Capt. William E. Fife, severely wounded at Cedar Creek; B. B. Sterret and Columbus McCoy, wounded at the first battle of Winchester; W. H. Peck, lost an arm in action at Piedmont; R. E. Bryan, hand shattered in action at Fayetteville; Casey D. Eskew and Robert Washington, taken prisoners and both died in confinement at Fort Delaware; Charles Bronough, John D. Wyatt, and Adison Newman, died in camp in the early part of the war; William Henson and T. H. Harvey, wounded in action at Fort Donalson; B. L. Hill, died of disease on New River; William Meeks, killed by a citizen of Buffalo, while on a visit home; William Morgan and W. H. George, killed in the Battle of Fayetteville; I. V. Newman and S. E. Staten, wounded in action at Cloyd's Mountain; John Farrow, killed in battle at Leetown, Virginia. In addition, a number were captured and served long terms of imprisonment.

So read the death rolls and records of battle-scarred veterans of the Kanawha Valley. Some wore blue, some wore grey, but the dead and the living died and suffered alike for what they believed to be right. They were soldiers in the full meaning of the term, and were descended from a pioneer ancestry of whom it was said: "These are farmers today, statesmen tomorrow, and soldiers always," and the performances of the Valley men in the Civil War lend honor to their ancestral heritage, and main the reputation of the soldiery of the Great Kanawha Valley. With the return of peace, these men came home, laid by the military trappings, donned the citizen's garb, and united in an effort to secure the intellectual and industrial development of their beautiful Valley.

A grim reminder of the Civil War's aftermath can be found today on a high overlook opposite the mouth of Coal River. There, in an old grown-over graveyard, you can find two tombstones, side by side, and almost leaning toward each other. On reading the inscription on the weather-beaten stone on the left you will discover that James Rust, a Confederate soldier, was killed in the Battle of Fayetteville; while the tablet on the right marks the grave of William Gregory, his brother-in-law was killed at the Battle of Lewisburg. (See page 112.)

Old-timers have passed the story down to us that these two young men had lived on adjoining farms and had been boyhood friends.

From *The History of the Great Kanawha Valley*, by John P. Hale and Virgil A. Lewis

---

ROLL OF BUFFALO GUARDS: Captain—Fife, Wm. E., (w), 2d Lieutenant—B.B. Sterrett, (w), 1st Sergeant—Wm. L. Bryan, 2d Sergeant—Wm. H. Peck (w), 1st Corpal [sic]—L. J. Timms, 2d Corpal—Thos. H. Harvey (w), 3rd Corpal—B.R. George. PRIVATES: Alexander, Sam'l T.; Bryan, Rees E. (w); Burch, Geo. W.; Bronaugh, Chas. E. (d); Brown, Wm. W.; Bailey, Geo. (s w); Bowles, Jerry; Brooks, Thos. W. (s), Byrne, R. V. (s); Craig, Clark; Claughton, Wm. F.; Collins, Jno. O. (s); Chapman, Wm. M. (s); Deem, Sam P.; Davis, Jno (s); Davis, Hen. C. (s); Eskew, Dorsal; Eskew, Casey; Farrer, Democracy; Fry, Wm. H.; Foresinger, Albert D. C. (s); Foresinger, George (s w); Giles, Sam W.; George, Andrew F.; Goings, Sam; Harmon, Edmond; Handley, Monroe; Horn, Christopher; Hogg, Henry H.; Henson, Wm. P.; Jones, Bird L. (k); Legue, Wm. G. (k); Legue, Simeon; Meeks, Wm. L.; Morgan, Wm. S. (k); McCoy, Sam A.; McCown, Henry M.; Newman, I.V.; Newmire, A. (d); Drummer—Neal, Hugh F.; Fifer—Neal, Jno. H.; Norvel, John (Snell); Peck, Isaac; Parmer, J. D. (s k); Rait, Thos. (s); Shank, Chas. E.; Sterrett, Sam A.; Smith, E.A.E.; Staton, Simon C. (w); Samuels, A.B. (s w); Samuels, Hugh (s w); Townsend, Paul (s); Wyatt, Jno. D. (d); Wood, Jno. H.; Woods, Joseph; Wallace, Luke T.; Watkins, Andrew J.; Withers, E.D.; Wells, Nicholas (s k)

k—killed, w—wounded, d—died, s—Sweeney's Company who disbanded and joined this company.

Baptism of Fire

1861 Campaign

RIPLEY, VIRGINIA,
July 6th, 1861.

To the true and loyal citizens of Western Virginia, and particularly those on the Ohio border, I would earnestly appeal to come to the defense of the Commonwealth, invaded and insulted as she is by a ruthless and unnatural enemy. None need be afraid that they will be held accountable for past opinions, votes, or acts under the delusions which have been practiced upon the Northwestern people, if they will now return to their patriotic duty and acknowledge their allegiance to Virginia and her Confederate States as their true and lawful sovereigns. You were Union men, so was I, and we had a right to be so until oppression and invasion and war drove us to the assertion of a second independence. The sovereign State proclaimed it by her Convention and by a majority of more than one hundred thousand votes at the polls. She has seceded from the old and formed a new Confederacy; she has commanded and we must obey her voice. I come to execute her commands, to hold out the olive branch, to the true and peaceful citizens, to repel invasion from abroad and subdue treason only at home. Come to the call of the country which owes you protection as her native sons!

HENRY A. WISE,
Brig. Gen'l.

John Letcher (1813–1884), CSA governor of Virginia during Civil War 1861–64. After the war he returned to his law office and dropped out of politics.

TO THE PEOPLE

Of the Department of the **KANAWHA VALLEY**, embracing the following Counties, viz: Mason, Jackson, Putnam, Cabell, Wayne, Logan, Kanawha, Boone, Wyoming, Raleigh, Fayette, Nicholas and Clay: According to the following order, by the

Governor of Virginia:

Executive Department, April 29, 1861.
LIEUT. COL. McCAUSLAND:

Sir: You will proceed at once to the Kanawha Valley and assume command of the volunteer forces in that section, and organize and muster the same into the service of the State; and as soon as they are formed into Battalions or Regiments, report the fact to me, with the names of the company officers, the number of men in each company, and the kind and quality of arms.

Gen. Lee will give all necessary orders for your government in that command. I am very respectfully,
JOHN LETCHER.

I have arrived here to take command of the Department. I have instructions to call into the field ten companies, and one company of artillery. These troops will be encamped in the Kanawha Valley, near Buffalo, Putnam Co. They are intended for the protection of the Department, and I appeal to the people of the border counties to abstain from anything which will arouse ill feeling on either side of the Ohio river. This Department is organized by the proper authority in the State, and is provided with the credit to sustain itself; but for complete success, I firmly rely on the friendly disposition of the people therein.

The volunteer companies of the counties of Mason, Jackson and Putnam, will rendezvous at **BUFFALO**, Putnam Co.

The volunteer companies of the counties of Cabell, Wayne, and Logan, will rendezvous at **BARBOURSVILLE**, Cabell county.

The volunteer companies of the counties of Kanawha, Boone, Wyoming, Raleigh, Fayette, Nicholas and Clay, will rendezvous at **CHARLESTON**, Kanawha county.

The Captain of the volunteer companies in the above counties will remain at their respective drill grounds, until ordered to their rendezvous by the Commandant of the Department. So soon as preparation to receive them can be made, the companies will be ordered to their respective rendezvous, mustered into the service of the State, and then ordered to the Camp of Instruction. No company will be mustered into service unless it has at least 82 men.

The Captains will see that each man is provided with a uniform, one blanket, one haversack, one extra pair of shoes, two flannel shirts (to be worn in the place of the ordinary shirts), two pairs of drawers, four pairs of woolen socks, four handkerchiefs, towels, one comb and brush and tooth-brush, two pairs white gloves, one pair of rough pantaloons for fatigue duty, needles, thread, wax, buttons, &c., in a small buckskin bag. The whole (excepting the blanket) will be placed in a bag, this bag will be placed on the blanket and rolled up, and be secured to the back of each man by two straps.

Lt. Col. JNO. McCAUSLAND,
Commanding Dep't Ka. Valley.

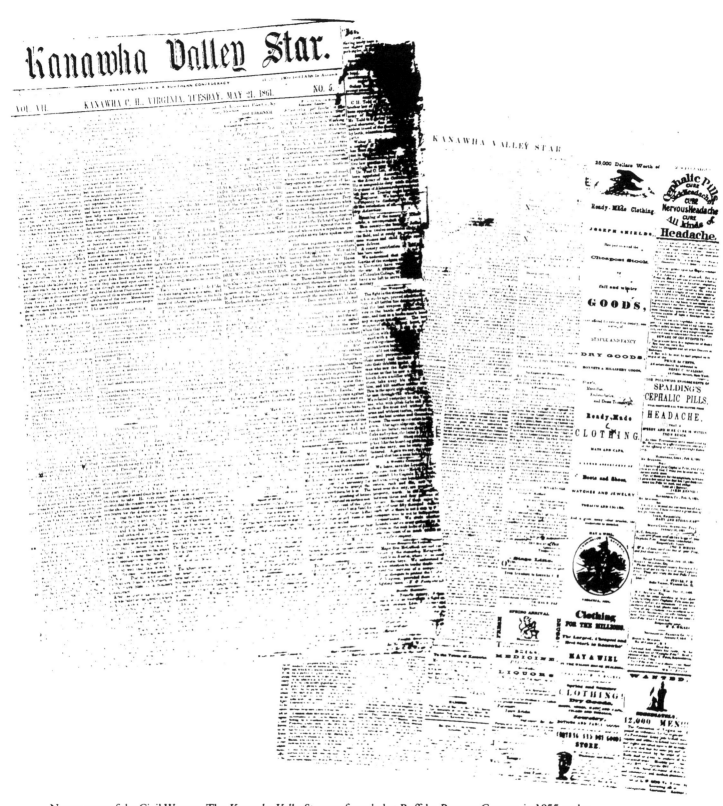

Newspapers of the Civil War era. The *Kanawha Valley Star* was founded at Buffalo, Putnam County, in 1855 as the *Star of the Kanawha Valley*. It moved its office to Charleston in 1857 and expounded the Democratic, pro-Southern cause. The publisher was John Rundle, who, as a member of the Kanawha Riflemen, joined the Confederate army in 1861. Union troops confiscated his press when they occupied Charleston in July 1861. Only two copies are known of *The Guerilla*, which was published in Charleston during the Confederacy's short occupation of the Kanawha Valley in September 1862. The *Charleston Daily Bulletin* was published in 1864 by E. T. and S. Spencer Moore and contained much war news and advertisements. The Moore Brothers were also job printers and dealers in newspapers, including *West Virginia Journal*, periodicals and stationery supplies. All newspapers are on file at the West Virginia State Archives.

Men of Virginia!
MEN OF KANAWHA!
TO ARMS!

The enemy has invaded your soil and threatens to overrun your country under the pretext of protection.

You cannot serve two masters. You have not the right to repudiate allegiance to your own State. Be not seduced by his sophistry or intimidated by his threats. Rise and strike for your firesides and altars. Repel the aggressors and preserve your honor and your rights. Rally in every neighborhood with or without arms. Organize and unite with the sons of the soil to defend it. Report yourselves without delay to those nearest to you in military position. Come to the aid of your fathers, brothers and comrades in arms at this place who are here for the protection of your mothers, wives and sisters. Let every man who would uphold his rights, turn out with such arms as he may get and drive the invader back.   C. Q. TOMPKINS,

Col. Va., Vol's. Comdg.

Charleston, Kanawha, May 30, 1861.

"Charleston is quite a pretty place."

GALLIPOLIS DISPATCH
August 8, 1861

The Commanders

Col. Christopher Q. Tompkins was a retired officer before the war, who had moved to the Gauley Bridge area and built a fine home. He had been a mining agent and superintendent of coal mines in Fayette and Kanawha counties and was 47 years old in 1861. He was commissioned a colonel of Virginia Volunteer forces in the Kanawha Valley, but resigned from the service on Nov. 6, 1861. VIRGINIA STATE LIBRARY

Upper right: Gen. Albert Gallatin Jenkins (1830–64), Confederate general who was born in Cabell County. He was a lawyer and representative to the Confederate Congress. He fought under Gen. John Floyd in the Kanawha Valley, led a raid through central West Virginia in 1862, and was killed at the Battle of Cloyd's Mountain in 1864. WVU

Gen. William Starke Rosecrans, a well-known Civil War officer in western Virginia and the western theater of operations, had business ties to Kanawha County prior to the war. A native of Ohio, he graduated from West Point and served in the U.S. Army from 1842–54. After resigning from the army in 1854, he went into the cannel coal business in the Coal River area. He also worked with other coal operations to improve navigation on the Kanawha River. In April 1861, Rosecrans rejoined the Union army and served as an aide-de-camp to Gen. George McClellan during the campaign in western Virginia. When McClellan was called to Washington to take over the Union command, Rosecrans was made brigadier-general and took over command in western Virginia. After April 1862, he commanded troops in the western theater with mixed results. Resigning from the army in 1867, he served as Minister to Mexico from 1868–69, then settled in California where he was elected to two terms in Congress from 1881–85, and then served as Register of the Treasury, 1885–93. He died in 1898. TERRY LOWRY

Gen. John Floyd (1806–63), Confederate general, former governor of Virginia and President Buchanan's Secretary of War. He was commander of forces under Gen. Robert E. Lee in his 1861 western Virginia campaign and commander at the Battle of Carnifex Ferry on Sept. 10, 1861. His feud with Gen. Henry Wise had a direct bearing on the failed Confederate campaign in the Kanawha Valley in 1861. WVSA

Upper right: Gen. Henry Wise (1806–76), Confederate general and former governor of Virginia. He was in command of his "Wise Legion" in the Kanawha Valley in July 1861 and under Gen. Robert E. Lee in his western Virginia campaign until relieved of command in September 1861. PHC

Gen. Jacob Cox (1828–1900), commander of the "Brigade of the Kanawha" at the beginning of the war and Federal commander at the Battle of Scary Creek. He was head of the Department of the Kanawha until August 1862. He was governor of Ohio, 1867–69, and Secretary of the Interior, 1869–70. USAMHI

Gen. John McCausland (1836–1927) raised Confederate troops in the Kanawha Valley in the early months of the war. He fought with Gen. John Floyd in his 1861 campaign and appeared in West Virginia throughout the war. He burned Chambersburg, Pa., in 1864 and refused to surrender at Appomattox. He spent the next two years traveling in Europe and Mexico before returning to the Kanawha Valley, where he lived until 1927, still insisting he was an "unreconstructed and unregenerate rebel."

WVSA

Dr. John P. Hale lived for 62 years in Kanawha County during an epoch of the country's most rapid development. Hale, the grandson of William Ingles and Mary Draper—the first white couple wedded west of the Alleghenies—was born in 1824 at Ingles Ferry in southwest Virginia. Mary Draper was the first white woman to see the Kanawha Valley, where Hale would move to in 1840. From 1841 to 1842 he attended Mercer Academy in Charleston, then studied medicine under Dr. Spicer Patrick. After graduating from the University of Pennsylvania Medical School in 1845 he formed a partnership with Dr. Patrick in Charleston.

Hale gave up medicine in 1847 and went into the booming salt business. By 1860 he had consolidated various properties in the Snow Hill area east of Charleston, creating the largest salt-producing business in the nation. Ironically, this was the same area where his grandmother, while a captive of the Shawnee Indians in 1755, had stopped with her captors when they made salt on their way west.

At the onset of the Civil War, Hale organized an artillery battery for service in the Confederate army and fought at the Battle of Scary Creek. He apparently left the service in 1863. In 1864 record show him back in the Kanawha Valley, president of the Kanawha Salt Company.

After the salt business collapsed in the 1870s, Hale pursued other interests. He introduced the first brickmaking machinery into the valley and financed the laying of the first brick street in Charleston in 1870. He helped organize the Bank of the West in 1863 and a gas company in Charleston in 1870. In 1864 he built the first steam packet boat on the upper Kanawha River, and in 1878 he constructed the steamers *Wild Goose* and *Lame Duck* using the first steam boiler made by the Ward Engineering Works. At one time he owned all the ferries in the Charleston area.

Hale was the man chiefly responsible for building the first state capitol building in Charleston. In 1871 he was elected mayor of Charleston. In 1872 he built his famous hostel, The Hale House. In that year he was president and part-owner of the company that started Charleston's first daily newspaper. And during this same era he established the first steam laundry in the city and organized the public delivery of ice.

In the 1880s Hale was engaged in the coal and timber business, but like his salt business, these ventures eventually failed, leaving him in dire financial straits. He always seemed to bounce back from his failures to build another fortune.

His contributions to the historical record are numerous. In 1883 he published a pamphlet on Daniel Boone's years in the Kanawha Valley, a chapter of Boone's biography that was practically unknown. In 1886 he produced his *Trans-Allegheny Pioneers* and in 1891 his *History of the Great Kanawha Valley*. He helped found the West Virginia Historical and Antiquarian Society in 1890. Hale was not a member of any established church and never married. He died in July 1902, and he is buried at Spring Hill Cemetery. His accomplishments were many, his life full, and his influence on the citizens of Kanawha County still felt.

The Home Front

What better way to make history come alive than to have the actual characters who lived it tell their own story? By referring to letters, diaries, and journals written by people who were in the Kanawha Valley during the Civil War, we can get a close-up and personal account of the way it was.

In early April 1861, as the clouds of war began to gather, Solomon Minsker, a Charleston miller, wrote his brother in Cumberland County, Pennsylvania, as follows:

Dear Brother:

The political excitements are running very high here now that the Virginia State Convention is in session in Richmond and it is feared that Virginia will go out of the Union. Most of the people do not like Lincoln's inaugural address but I hope all will be settled without coming to arms. The South, however, is preparing for it. We have two military companies in this town now fully uniformed and equipped. I belong to one of them.

Victoria Hansford was a young woman living in Coalsmouth (now St. Albans) at the beginning of the war. She wrote the following in her journal "The Reminiscences of the War—1861."

The spring of 1861 came as all other springs with its sunshine, its birds, and its flowers, yet we hailed it not with the usual joy and anticipation. Far away we could hear the rumbling of the storm that we feared would soon sweep over our beautiful valley. Fort Sumter had already fallen, there had been blood shed in Maryland, and now there had been a great stir in Western Virginia.

Captain George Patton from Charleston had heretofore been a staunch Union man and had previously made stirring speeches in our town in favor of the Union, but recently, however, he returned calling on all sons of Virginia to rally around her flag. He was cheered lustily and how patriotic we felt as we waved our handkerchiefs. We all sang "We Will Die for Old Virginia." Altogether, this was an exciting day in our quiet little village of Coalsmouth.

Nine of our young men volunteered in Captain Patton's company known as the "Kanawha Riflemen." They were: my brother Carroll Hansford, Stephen Teays, N. B. Brooks, Charlie Turner, Theodore Turner, Thornton Thompson, Tom Grant, James Rust and Henry Gregory. The citizens of Coalsmouth gathered around the wagon to tell them good-bye and to wish them God speed. Tears were not only shed by their mothers and sisters, but many others that day wept over the sacrifices they were about to make.

The concerns and fears of those gathered there that day were justified since six of the nine young volunteers would die in service and the other three would be captured and left to languish in northern prisons.

The war broke out on April 12, 1861, and five days later Virginia seceded from the Union.

On May 11, 1861, Lt. Col. John McCausland published a recruitment ad in the *Kanawha Valley Star.* It read:

I have arrived to take command of the Department of the Kanawha. I have instructions to call into the field ten companies of infantry and one company of artillery. The captains of these volunteer companies will see that each man is provided a uniform, one blanket, and haversack, one extra pair of shoes, two flannel shirts, one comb and brush, toothbrush, two pairs of drawers, four pairs of woolen socks, four handkerchiefs, two pairs of white gloves, one pair of rough pantaloons, needles, thread, wax and buttons, secured in a small buckskin bag. This all will be placed in a bag that will be rolled up in a blanket and secured to the back of each man by two straps.

Scary Creek

By early July 1861 Union General Jacob D. Cox had launched a three-pronged attack on the Kanawha Valley. His main forces were loaded on four steamboats and were proceeding up the Kanawha River. Later he wrote in his "Military Reminiscences" his impressions of their first day in enemy territory.

Our first days sail was thirteen miles up the river, and it was the very romance of campaigning. I took my station on top of the pilot house of the lead boat, so that I might see over the banks of the stream and across the bottom lands which bounded the valley. The afternoon was a lovely one. Summer clouds lazily drifted across the sky while the boats were dressed in their colors and swarmed with men as a hive of bees. The bands played national tunes, and when we passed the houses of loyal Union citizens, the people would wave their handkerchiefs and they were answered by cheers from the troops.

The scenery was picturesque, the gently winding river making beautiful reaches that opened new scenes at every turn. Now and then a horseman would bring a message to the shore from the advanced guard on either side and a small boat would be sent out to get it. It gave us just enough excitement and the spice of possible danger to make the first day in the enemy's country key everybody up and doubled the awareness of every sensation. The landscape seemed more beautiful, the sunshine more bright, and the exhilaration of outdoor life more joyous than any we had ever known.

When news of Cox's invasion of the Kanawha Valley reached Coalsmouth where the Confederate Camp Tompkins was located, it caused a flurry of activity among the noncombatants left on the home front. Since it was yet a week before the first Battle of Bull Run, they were literally seeing the Civil War erupt in their own front yard. Victoria Hansford was there and she wrote about how families reacted and did the best they knew how to prepare for war.

My brothers would not hear to my remaining home any longer as I was already the only white female still in town. They directed that I should refugee to Paint Creek and stay with my Uncle Felix Hansford until after the battle. Ah, those were heart rendering times. It was sad to go away and leave my father and two brothers behind with the enemy advancing up the valley. The servants, six in number, were sent a few miles up Coal River to Frank Thompson's where he had quarters for them.

People went in all directions to get out of the village as we thought the Yankees would surely burn our homes if they got this far, since the soldiers had been encamped here. My brother Charley gave me his money to keep and I dug a hole and buried it with all my mother's silver and other valuables.

I left home the day before the Battle of Scary and we started for Paint Creek. The road was full of refugees going up the valley while soldiers and armed civilians were going down the valley toward the advancing foe. Weeping women and unhappy children were all along the road. When I got to my Uncle Felix's house at the mouth of Paint Creek, I found many others in the same fix as I was.

Sarah Frances Young was nineteen years old in 1861 when the Civil War brought many changes to her young life. She lived on a small farm near Coal Mountain in Putnam County and she wrote in her "Little Journal" how the war affected her and her family. Her father, Capt. John Valley Young, was Company Commander of Co. G, 13th Regiment, Virginia Volunteer Infantry, of the Union Army. After the Bat-

tle of Scary he was left in charge of the occupation forces at Coalsmouth, now St. Albans. Since the Youngs were active in the Union cause, while most of their neighbors were pro-Confederates, they were subjected to many incidents of harassment and oppression, which she wrote about in her journal.

July 13, 1861
Oh, what a time the Federal troops are coming up the Kanawha. The Seceshionists are running with their guns to bushwack them. They are mad because Father won't raise arms against our government. So, we are glad to see our protectors come. Gen. Cox has landed at Winfield, and will move on up in a few days.

July 17th, 1861
Two Rebels came by today begging for something to eat. They had been gone but an hour before the Federals attacked the Rebels at the Mouth of Scary. Oh, my! How our hearts ached as the sound of the booming cannons reached our ears. We could distinctly hear the fire of the small arms. We were afraid the Federals would be whipped but now it is night and we have heard the result. Three Rebels were killed and eight Federals. Col. Norton, with several others, were captured. May they escape soon and rejoin their regiments.

Sgt. James H. Collins, a Union soldier from Buffalo, told about his part in the action at Scary Creek:

We marched down the road with our guns on our shoulders until the sound of the enemy's small arms fire warned us we were coming under fire. The mouth of Scary at that time was the best natural defensive position along the Kanawha. The creek passes through a deep ravine with high, steep and rugged banks. On reaching the creek, it was obvious we could not cross at that point. When an order was given to withdraw, we thought it meant retreat, so we all skedaddled to the rear.

From the Confederate side of the line Kanawha Rifleman Levi Welch, from Charleston, also told about his part in the action at Scary Creek:

Capt. Albert G. Jenkins, later Brigadier General, came up the skirmish line, with his hat off and blood streaming down his hair and neck. He called for someone to go get his horse that was staked out behind Hale's battery. So I left my place behind a beach tree and ran back over the hill and got his horse.

I rode back to the battery and saw a disabled cannon being propped up by a bunch of determined men. I asked them where is my brother?

Who is your brother, they answered. Lt. James Welch of this battery. Over there he lies, he has done his duty. I looked where the soldiers pointed and saw my brother upon the ground lying where he fell, with his head almost severed by a piece of flying iron. It was from the carriage of the gun he was aiming when it was struck by enemy artillery. As he lay with both arms extended in the form of a cross, he reminded me of Christ being crucified. One died for all mankind, the other for his native state, with the same willingness.

Sarah Young noted:

July 24th, 1861
General Wise has left the Kanawha Valley. I hope he may never see Kanawha again. The Federals now have possession of Charleston and they are received by the Union men with great joy.

An item in John Morgan's "Charleston 175" refers to a letter written by a Union soldier in Cox's invading force that took control of Charleston after Wise's retreat.

Camp Norton, Aug. 5, 1861
. . . Embarked on the Silver Lake No. 2 and after a delightful trip up the beautiful valley of the Kanawha arrived at this place at 7 o'clock in the evening. At every house and landing along the river the boys met the heartfelt cheers of welcome from men, women and children . . .

Camp Norton is situated on the grounds of the court house—headquarters being in the building. Capt. J. L. Vance is commander of the post.

The officers keep strict guard over the men . . . the citizens of Charleston have nothing to fear from them.

Charleston is quite a pretty place, with about 2,000 inhabitants and is located on the beautiful bottom of the north-east bank of the river and surrounded by lofty hills.

There are many pretty residences, but they and the public buildings are built after the old style and have not much pretension to magnificence. Peace reigns in the vicinity once more and business is resumed.

The Bowen Tavern (Six Mile House) at Tyler Mountain. Confederate soldiers from General Wise's command used the water from the tavern's well in the summer of 1861. The house is no longer standing.

Littlepage Mansion, on West Washington Street at Littlepage Terrace, Charleston. Built in 1845 by Adam Littlepage, it sat at the junction of the Ripley-Ravenswood Road and the Point Pleasant Road. In July 1861 the area was the headquarters of Confederate Gen. Henry Wise, who established a camp around the house. When Mrs. Rebecca Littlepage denied him entrance, Wise threatened to blow up the residence with a cannon. Today it is used as an office by the Charleston Housing Authority.

Site of Camp Two-Mile on Littlepage farm, Charleston, from a 1901 photo. PHC

Site of Camp Tompkins, established on the farm of William and Beverly Tompkins. Col. C. Q. Tompkins selected the site because of its strategic location and to give reassurance to residents of the area by hindering Union sentiment there. It was located slightly west of the mouth of Coal River near Tackett's Creek and the small community of Coalsmouth. The camp became the center of Confederate activity as recruits poured in on a daily basis in the early summer of 1861. Col. C. Q. Tompkins (no relation to William Tompkins) and Captain Patton made their headquarters in Tompkins home called "Valcoulon Place." The site is now occupied by the Valley Drive-In theater.

The Littlepage Stone Mansion was built in 1845 by Adam Littlepage. Today it is used as an office for the Charleston Housing Authority.
COURTESY MRS. JOHN T. MORGAN

In 1861 Mrs. Rebecca Littlepage, mistress of the Littlepage Mansion, denied entrance to the house by General Wise, who threatened to blow it up by cannon fire.
COURTESY MRS. JOHN T. MORGAN

A post-war view of the apple orchard at Camp Two-Mile. The Confederates established the camp on the farm of Adam Brown Dickinson Littlepage, along Two-Mile Creek. The military camp extended from the Kanawha River to the junction of the Ripley-Ravenswood road and the Point Pleasant road (present West Washington Street). It is now the site of the Orchard Manor housing development. COURTESY MRS. JOHN T. MORGAN

Valcoulon, an antebellum mansion on the farm of William and Beverly Tompkins at Coalsmouth, was the headquarters for Col. George S. Patton before the Battle of Scary Creek. It is on the presentday Valley Drive-In theater, just off Route 35 and 60 near St. Albans. GARLAND ELLIS

Pryce Lewis, a Union spy employed by Alan Pinkerton, spent time in the Kanawha Valley prior to the Scary Creek battle obtaining information on Confederate troops and their fortification. He posed as an English tourist and was befriended by Colonel Patton and Colonel Tompkins, but was rebuffed by Wise when he tried to get a pass to travel to Richmond. Although he eventually made it back to Union lines in Ohio, he was too late to give any useful information to General Cox, who had already fought at Scary Creek. MARY LEWIS COLLEGE

Pryce Lewis, glass in hand, toasts Col. George S. Patton during their meeting at the Valcoulon Mansion, Camp Tompkins. Sam Bridgeman, Lewis' assistant and traveling companion, looks on from the left.
FROM ALAN PINKERTON'S *SPY OF THE REBELLION*

Early Campaign Correspondence
of Lt. Col. John McCausland

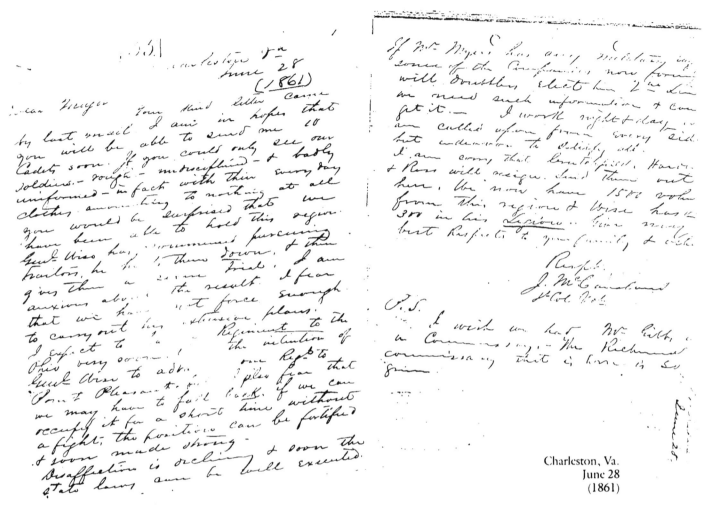

Charleston, Va.
June 28
(1861)

Dear Major—

Your kind letter came by last mail. I am in hopes that you will be able to send me 10 Cadets soon. If you could only see our soldiers—rough—undisciplined—& badly uniformed—in fact with their everyday clothes, amounting to nothing at all. You would be surprised that we have been able to hold this region. Genl. Wise has commenced pursuing traitors, he hunts them down and then gives them a severe trial. I am anxious about the result. I fear that we have not force enough to carry out his extensive plans. I expect to lead a Regiment to the Ohio very soon. It is the intention of Genl. Wise to advance one Regt. to Point Pleasant but I also fear that we may have to fall back. If we can occupy it for a short time without a fight, the position can be fortified and soon made strong.

Disaffection is declining & soon the state laws can be well executed. If Mr. Myers has any military information of the companies now forming will doubtless elect him 2nd Lieut. We need such information & cannot get it. I work night and day. I am called upon from every side, but endeavor to satisfy all.

I am sorry that Crutchfield, Hardin & Ross will resign. Send them out here. We now have 1500 volunteers from this region & Wise has only 300 in his Legion. Give my best Respects to your family & others.

Respt.
J. McCausland
Lt. Col. Vol.

PS—I wish you had Mr. Gibbs as a Commisary. The Richmond commisary that is here is so grimm.

THIS LETTER AND THOSE ON FACING PAGE ARE FROM THE VIRGINIA MILITARY INSTITUTE ARCHIVES.

TOP LEFT, FACING PAGE

Dear Major

I have charge of a large encampment at this place. I need at least 15 good Instructors. Please send me 15 Cadets or anyone capable of instructing. We expect to have a fight soon, in fact the Ohio troops have advanced within 50 miles of this place & have orders to advance as soon as they are reinforced. Gov. Wise is expected with his Brigade. I'm afraid that he will not arrive in time. If you can accomodate me I wish you would.

Headquarters
2 Mile Camp
Near Charleston, Va.

Truly yours
J. McCausland
Lt. Col. Va. Vol.

Head Quarters
Buffalo
May 16 (1861)

Dear Col.
    I am in need of drill masters. Send me some. I am preparing rapidly. I sent to the Col. of Ordnance for 500 cartridge boxes, 500 bayonets, scabbards, if they have not been sent please send them & send some fixed ammunition. Have you been watching the Wheeling Convention, they will apply to Lincoln for arms & men & will resist authority of the state. Enclosed I send a letter I received. I receive daily such letters and influential men tell me the same thing. The Ohio paper suggests that the men I now have should be captured, as the Missouri troops were. I have only to say they can't take one of us alive. Please send us tents as our present camp cannot be moved. I have not yet made my return to the Governor Inspector Genl. I will do it soon. I labor very hard & will try to give a good account of my conduct.
    Could we get 5000 foot troops as the war may begin here.

                Respectfully
                J. McCausland

Charleston, Va.
July 3rd

Dear Major
    Please send me as many cartridges for flint muskets as you can spare. Powder is very scarce & difficult to get. Cadet Thompson arrived today, much obliged to you for him. Please send the rest Chen, Rouse, Beck, Kin & others.

                Respect.
                J. McCausland
                Col. commanding
                Regiment Va. Vol.

Notes, Quotes & Anecdotes on the Battle of Scary

The Battle of Scary Creek had little if any effect on the outcome of the Civil War; however, it still remains one of the most outstanding historic events that ever took place in the lower Kanawha Valley.

Scary became an epic in the annals of our local history because it involved, in some way, practically every person living in the area. Most of the Confederate forces were native sons, and they lived to tell over and over again the stories of their own personal experiences during the battle. Hundreds of people living in the valley could also remember seeing troop movements to and from the battle, and many even heard the cannon's boom and the rattle of musket fire. A collection of these personal recollections have provided a storehouse of human interest stories which were much more exciting than the official military records.

The Battle of Scary was unique because it was the baptism under fire for most of the troops involved. Both men and horses lost their composure when they were first exposed to the shock of battle. Poor training, lack of discipline, and inexperienced officers were some of the other reasons that converted the battlefield to a scene of disorganized mass confusion.

The prelude to the Battle of Scary began when Gen. Jacob D. Cox crossed the Ohio River at Pt. Pleasant early in July 1861, with orders to drive General Wise out of the Kanawha Valley. He had, under his command, about 3,000 raw recruits from the areas around Cincinnati and Columbus, and the majority had less than two months' service. Cox had to transport most of his troops up the Kanawha on four steamboats, because two days before he had marched them approximately 25 miles to Gallipolis, and their feet had become too sore for him to walk them any farther. The trip up the river was a gala affair with flags flying and bands playing.

The seriousness of war was realized by Cox, however, two nights later when his troops had reached the vicinity of Winfield. While pickets were being posted in the dark on the south side of the river, a soldier accidentally discharged his gun. This caused the other trigger-happy recruits to begin firing in all directions at the enemy who was not there. When the firing was finally stopped, it was discovered that two young soldiers had been killed.

Meanwhile, the Confederates in the Kanawha Valley had been much more serious in their preparations against the invasion of their homeland. Around the middle of May 1861, Col. Christopher Tompkins had established a camp on the farm of William Tompkins, near the mouth of Tacketts Creek below St. Albans. This was to be the collecting and training point for all volunteers from the area. Col. John McCausland had also set up headquarters at Buffalo, where he began gathering recruits to be funneled into the units being organized at Camp Tompkins.

All the young men in the upper class at Buffalo Academy volunteered in a body, and most of them took part in the Battle of Scary.

On June 26, Gen. Henry A. Wise arrived in Charleston with the almost impossible orders to raise a combat-ready brigade from raw untrained volunteers. With no arms, supplies, or transportation furnished, it was a wonder that his forces were able to make any kind of a stand at all in the Kanawha Valley. To make matters worse, Wise had no military training or experience, and he was erratic, impulsive, explosive, and thoroughly unpredictable. It has been said that the best thing that happened to the Confederate troops the day of the battle, was the fact that General Wise was not present.

Volunteers from local militia units poured into Camp Tompkins, with each man being required to bring his own arms and equipment. They carried nearly every type of firearms—squirrel rifles, shotguns, flintlocks, and an assortment of belt guns. Only two or three units wore any sort of uniforms, and the men in the cavalry were required to furnish their own horses. It was a military miracle that such a motley aggregation could be molded into any kind of an effective fighting force in such a short time.

What the Confederates lacked in organization, however, they made up in military strategy. By destroying the bridge over Poca River they had halted Cox's advance on the north side of the Kanawha. They then chose to make their stand on the south side, at the mouth of Scary, where the hills came nearest to the river and the creek was too deep to ford. Johnson Shoals was also located in the river at this point, and the steamboat chute was on the south side in easy range of their guns. Thus, they had temporarily halted the Union advance up the valley by land and water.

The outcome of the battle is now a matter of

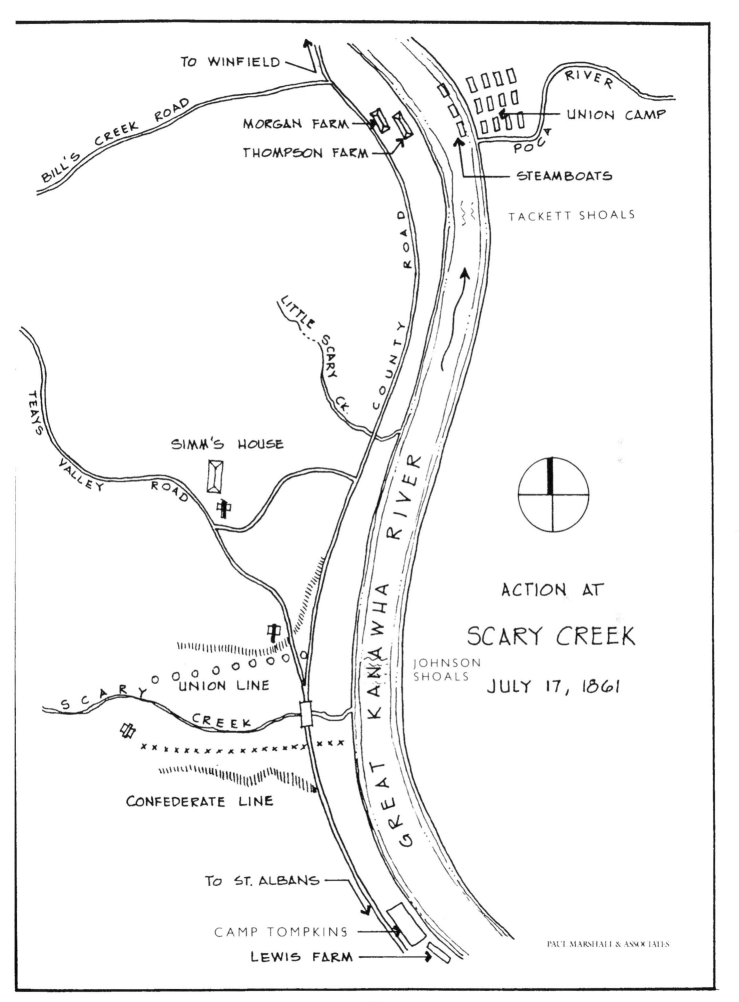

TO WINFIELD

BILL'S CREEK ROAD

MORGAN FARM

THOMPSON FARM

RIVER

UNION CAMP

POCA

STEAMBOATS

TACKETT SHOALS

COUNTY ROAD

LITTLE SCARY CK.

TEAYS VALLEY ROAD

SIMM'S HOUSE

GREAT KANAWHA RIVER

ACTION AT

SCARY CREEK

JULY 17, 1861

JOHNSON SHOALS

SCARY CREEK

UNION LINE

CONFEDERATE LINE

TO ST. ALBANS

CAMP TOMPKINS

LEWIS FARM

PAUL MARSHALL & ASSOCIATES

- 55 -

record in the Official History of the Civil War. In short, elements of the 12th and 21st Ohio Regiments attacked the Confederate position in the afternoon of July 17, 1861. After about five hours of attempting to dislodge the Confederates with frontal and flanking attacks, the Federals finally withdrew, after losing 15 killed, 11 wounded, and 7 captured. The Confederate losses were much lighter, with 4 killed, 10 wounded, and 2 captured.

The report of the battle went almost unnoticed in Washington, since it reached Union headquarters on the same day that the first Battle of Bull Run took place.

Even with his success at Scary Creek, Wise had become aware of his hopeless position in the Kanawha Valley. With Rosecrans marching toward Gauley Bridge by way of Weston and Sutton, and Cox moving on Charleston, he was in danger of being cut off in the valley by a superior force. His only alternative was to pull out, and without delay, on July 24th, he began his "Great Skedaddle" up the valley, which did not stop until he had reached Lewisburg.

After the battle, the wounded from both sides were brought back to the William Thompkins house at Camp Thompkins, and it was then converted to a hospital.

It was ironic that this historic old house should finally fall victim to still another war. In 1918 it was torn down to make way for the Rossler and Hasslacher Chemical Company, a World War I installation, which was one of the first plants in the world to produce chlorine gas by the electrolysis method.

When Captain Patton was made commander of the Confederate forces at the Battle of Scary, he was instructed by Wise not to make a stand at Scary, but to retire gradually to Bunker Hill near Upton Creek, concentrating his forces there. When the Union troops approached, however, Patton realized that Cox's forces were divided, and he quickly ordered his 800 men back to Scary Creek to meet the 1,200 attacking Federals.

During one of the Union charges, about three-fourths of the Confederates became panic stricken and began falling back in complete disorder. Patton mounted his horse and attempted to get the men back on the line. His already frightened mount, however, bolted toward the rear, and the men, thinking he was also pulling out, began to flee all the faster. He soon managed to get his horse under control, and was finally able to rally his troops and get them back on the line. In doing so, however, he became exposed to enemy fire and was shot through the shoulder.

Lieutenant Colonel Patton later wrote a letter to his brother in which he described his version of the battle. He also said: "It is now forgotten except by those who first met the enemy and still talk of Scary around their camp fires." This letter was preserved in the Patton household where young George S. Patton III grew up, looking to his grandfather as his idol and inspiration. If the Battle of Scary served no other purpose than to motivate the military zeal of the great World War II general, then it truly made an outstanding contribution to our national heritage.

Capt. Albert Gallatin Jenkins, who commanded the cavalry, assumed command when Patton was wounded. Captain Jenkins was from Cabell County, and later in the war he reached the rank of Brigadier General but died of wounds received at Cloyd's Mountain.

Jenkins was a namesake of Albert Gallatin, a Swiss immigrant who became a noted statesman and a close friend of George Washington. It was a coincidence that the first stream above Scary Creek, on the Kanawha, is known as Gallatin's Branch. It was so named when Albert Gallatin got his boat stuck on a sandbar at that place while on an exploration trip down the river in 1785.

A short time after Captain Jenkins assumed command, he received a scalp wound that caused the blood to stream down his face and neck. A young private named Levi Welch volunteered to go get his horse so he could better traverse his line of defense. On returning with the frightened steed, Welch could not remove his foot from the stirrup, and he was compelled to cut the strap. When he got in the saddle, Captain Jenkins was unable to control the horse and it bolted toward the rear. The men, on seeing their bloody captain go galloping away from the battleline, once again took off for the rear. Col. Frank Anderson, a member of General Wise's staff, was arriving from Charleston about the same time, and he was able to assist Captain Jenkins in getting the troops back on the line, once again.

Levi Welch returned to his place on the left end of the line just as the Federals were launching a determined charge to cross the bridge. About the same time, an unfamiliar force of red-shirted riflemen pressed in on his position. One soldier was shot before it was learned that they were friendly troops, which turned out to be the Sandy Rangers from Wayne County, under Capt. James Corns. The Blood Tubs, as they were called because of their red shirts, swung into action, singing "Bullets and Steel," which could be heard above the roar of the battle. This rallied the

Confederates and helped turn the tide of battle, causing the Union forces to abandon the field and to withdraw back down the river. One of the officers leading a company of the Sandy Rangers was destined to become an almost legendary figure in West Virginia history, known as "Devil Anse" Hatfield.

Levi Welch also recalled that later about dark, his company was burning some buildings so they could not be used for shelter in case the enemy returned. It was then that a group of Union officers came up to the fire, thinking they were in the midst of their own victorious troops. They were quietly surrounded and captured, however, before they had realized their mistake. It was a good haul; Col. William E. Woodruff, Col. Charles DeVilliers, and Lt. Col. George W. Neff were taken. These officers had crossed the river from Cox's headquarters near Poca, and it points out the lack of Union communications maintained during the battle.

Levi was the brother of Lt. James Welch, who was killed when the cannon he was aiming received a direct hit from the Union artillery. The next day young Welch received permission from Lt. Nicholas Fitzhugh to go to Upper Falls on Coal River where his mother was staying, to tell her of his brother's death.

Scary Creek was a costly engagement for Cox's army, considering the loss of field grade officers. Capt. Thomas J. Allen was mortally wounded; Col. Jesse Norton was severely wounded and captured; Colonel DeVilliers, Colonel Woodruff, and Colonel Neff were also captured. This left only Lt. Col. Carr B. White and Col. John W. Lowe of the 12th Ohio remaining in Cox's command. Colonel Lowe was also lost six weeks later when he was killed at the Battle of Carnifex Ferry.

Directly across the Kanawha River from the battle site lived the Mason family, whose father was an overseer for the Lewis Bowling plantation. Aside from being in danger of stray bullets, the family had a far greater concern during the battle. Two of their brothers, Thomas L. and William L. Mason were fighting in the Confederate ranks.

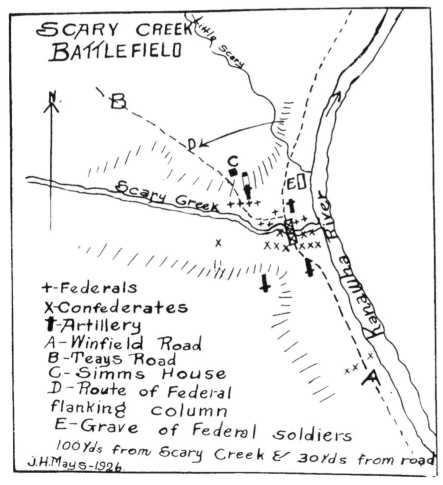

SCARY CREEK BATTLEFIELD MAP—As drawn by Confederate veteran James H. Mays in 1926.

1st Lt. William Anderson "Devil Anse" Hatfield, famous leader of his family in the postwar Hatfield and McCoy feud in southern West Virginia. He was reported to have been present at the Battle of Scary Creek. WVSA

Mae Teays, who later lived in St. Albans, recalled the family telling of how her two uncles had come across the river the night before the battle to check on the safety of the ones left at home. Ella Mason, Mae Teays' aunt, often told her how the children huddled behind heavy furniture during the height of the action. Both of the Mason brothers survived the battle only to be struck down later in the war. Thomas was killed at Lewisburg in 1862, and William fell at Leestown in 1864.

Only recently on the site of the Scary battlefield, another forgotten story was literally resurrected almost as if it had been told by a voice from the past. It all started while two boys were fishing in Scary Creek. Running out of bait, one of the boys began digging around for worms. Uncovering a small corroded piece of brass metal, he later learned he had found a Union officer's belt buckle.

Some time later Ed Gardner, a gun collector of Poca, came by the buckle through a gun trade. On examination, he discovered engraved on the back, the initials "B.B.B." and also a faint outline of the French fleur-de-lis, usually considered the national emblem of France.

By checking the muster rolls of the 12th and 21st Ohio regiments, only one soldier was found with the initials B.B.B. He was a private named Benjamin B. Bonneville. He volunteered for three years, along with his brother, and they both served their full term with Co. C, 12th Ohio Regiment without becoming a casualty.

It was later learned that Private Bonneville had an uncle named Benjamin L. Bonneville who was a Brigadier General. General Bonneville was born in France in 1795 and was brought to America by his parents to escape a purge by Napoleon. He graduated from West Point in 1815 and was eventually assigned by the Army to lead an exploring and mapping expedition into the far West. The Bonneville salt flats in Utah were first discovered and mapped by him.

When the French General LaFayette returned to America, Bonneville was assigned as his aide and interpreter and was with him continually while he was in this country. When the Civil War broke out, he was too old for field duty and was assigned as commandant of an army installation near St. Louis, Missouri.

Since Benjamin B. Bonneville, the private, had an officer's belt buckle, it is natural to assume that he had acquired it from his uncle, the General. The engraving of the French emblem on the back also sug-

gests the French background of Gen. Benjamin L. Bonneville.

Virgil A. Lewis said it best in his *History of the Great Kanawha Valley*: "Some wore the blue and some wore the grey, but they died and suffered alike for what they believed to be right. They were soldiers in the full meaning of the term, and descended from a pioneer ancestry of whom it was said: 'They are farmers to-day, statesmen to-morrow, and soldiers always.' Their performances in the Civil War lent honor to their ancestral heritage. With the return of peace these men came home, laid by their military trappings, donned citizens garb, and united in an effort to secure the intellectual and industrial development of their beautiful valley. How well they succeeded will be left to future generations to decide."

On Monday night the Telegraph No. 4, took on board a few men of the 41st, and paid a visit to the farm of one Albert Gallatin Jenkins, some twenty miles below this place. This man Jenkins has rendered himself notorious for his guerrilla feats of warfare, and as infamous as he is notorious.— He it was, who in connection with the villainous Clarkson, attacked Guyandotte, and refused quarters to all asking it. It is fit that he should thus be made to feel the same punishment he has been so ready to inflict on others. The boat returned on Tuesday about noon, with several hundred bushels of corn, 150 head of hogs, &c. &c. It is high time some such lessons should be administered to these scoundrels, who, not only commit every crime known to the decalogue, but violate every rule of civilized warfare.

From the Nov. 19, 1861, issue of *The Gallipolis Journal* concerning the "traitorous Virginian, Albert Gallatin Jenkins of Cabell County." The 41st was the 41st O.V. Regiment.

The question now arises as to what took place at Scary after the retreat of the Federal army. It has been the assumption by many that the bodies of the Federal soldiers were sent into the camp at Poca, and forwarded to Ohio. This now appears not to be the case. James H. Mays, of Company F, 22nd Virginia Infantry, in the *Charleston Mail*, records that his company arrived late, returning from a scout in another direction. "We could plainly hear the rattle and roar of guns and see above the tree tops the smoke of battle," he wrote, "and I dare say that the sight of that smoke and the sound of those guns caused more fear and trembling than any of the succeeding battles of the four years of the war." A messenger urged haste but when the company arrived the Federals had returned.

"The next morning (July 18th) continued Mr. Mays, "we went over the battle grounds and gathered up the dead, in order to bury the bodies. We found 14 Yankees which we dragged together. We dug a pit about six feet by twenty feet and spread some straw on the bottom . . . it was the best we could do under the circumstances. While dragging the bodies I had carried one of the men's caps. The name on the cap, I noticed, was 'Captain Allen.'" The cap undoubtedly belonged to Captain Thomas Allen, whose death has already been recounted. The place of burial was a short distance below the present junction of the Teays and Winfield roads, and its subsequent history is unknown.

Local inquiries bring a note that after the war a number of bodies were taken up near the mouth of Coal River and removed to the National Cemetery, at Grafton. Another account relates that about 1868 army men appeared at the sites, made excavations, but found no remains. The whole affair is not clear from any available records, but if the boys in Blue still sleep there, it would be little enough honor that the spots be marked.

—Roy Bird Cook

Unless you slow down and read the historical highway marker, you would never dream that over 2,000 brave men fought a battle to the death across this little creek. These 1995 photos tell us what has survived from that hot summer day in 1861. The greatest change was the huge C&O Railroad grade that obliterated much of the battlefield. Even with the 20th century's changes there is little doubt Lt. James Welch or Captain Patton would mistake this fateful place where one was to die and the other was badly wounded.

Monument at the bridge across Scary Creek. It was erected by the St. Albans Chapter of the United Daughters of the Confederacy to commemorate the battle fought here on July 17, 1861.

"When Cox's army was advancing upon Charleston many citizens in their fear of the Federals took themselves and their families to Cemetery Hill and one woman removed her petticoat and furiously waved it as a flag of truce."

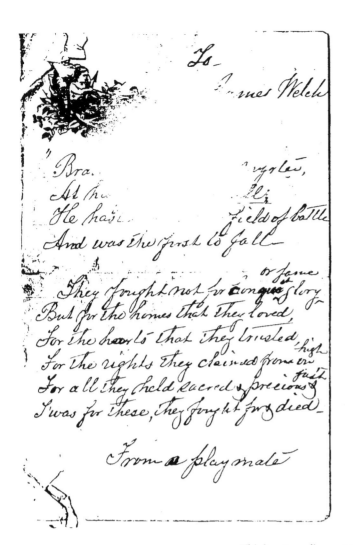

"Bra...   ...rgrles,
At the...   ...li,
He has...   field of battle
And was the first to fall

                                        or fame
They fought not for conquest glory
But for the homes that they loved,
For the hearts that they trusted,
                                              high
For the rights they claimed from on
                                              faith
For all they held sacred & precious &
I was for these, they fought & died.

                    From a playmate

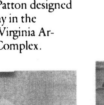

Lt. James Clark Welch, C.S.A.

A wonderful relic of the Kanawha Riflemen—this dark green overcoat was a part of the splendid uniform that Capt. George S. Patton designed for his men. On display in the Museum of the West Virginia Archives in the Capitol Complex.

This heart-rending note from a boyhood playmate of "Jimmie" Welch written to Welch's mother resides in the West Virginia Archives.

Carved on his tombstone in Charleston's Spring Hill Cemetery is an epitaph that perhaps sums up the Civil War in fewer words than any other. Lt. James Clark Welch died on July 17, 1861, at Scary Creek. One of the first Confederate soldiers to die in the Kanawha Valley. Although the rains of over a century have eroded the letters, they still come forth in word's that probably his mother chose: "Died in defence of his native state."

To a student of Civil War history to be able to touch this appointment to 2nd Lt. James C. Welch is indeed memorable. The actual certificate is in the collection of the West Virginia Archives in Charleston, where we made this copy for our readers. Like most old cotton-content paperwork, it is in remarkably good condition. What a story it could tell—first Jimmie's pride in holding it and later his mother's tears as she realized her boy had fallen.

Headquarters Camp
Tompkins
June 16, 1861

Lt. Welch
     If the dredge boats come down the river tonight or today—you will *not* interfere with them, but allow them to pass.
     Has your guard heard any firing of guns, or cannon during the night—Answer at once.

Yours
Geo. Patton
Comm.

Special Orders
No.

Headquarters
Camp Tompkins
June 14, 1861

Lieut. Welch,
Sir,
     You will see to it that the dredge boats, now in Kanawha River, be not taken away; they must not be allowed to pass down, and if it be necessary, you must use your gun to stop them.

By order of
Capt. Patton, comm.
D. L. Ruffner
adjutant

P.S. Keep a bright lookout tonight for them.

- 61 -

The Battle of Scary
as told by Confederate Soldier Levi Welch

This first-hand account of the Battle of Scary was taken from a letter from Levi Welch to Dr. John P. Hale, for publication in the W. Va. Historical Magazine. Levi Welch was a Confederate soldier and took part in the battle. His brother. Lt. James Welch, was killed when the Federal troops scored a direct hit on the artillery piece he was in the process of sighting.

Dear Sir:

In compliance with your request, I hereby send you some of my recollections of the fight at Scary Creek, in this county, one of the first fights of the late Civil War.

The "Kanawha Riflemen" on July 17, 1861, were ordered from Camp Tompkins, near Coal's Mouth to meet the advance of General Cox's Federal forces. Being a private in this company, my observations were confined, most of the time, to a small field of vision.

The Riflemen were deployed as skirmishers, in advance of the other troops, in front of Hale's Battery extending up the ravine along a brush fence. I was the last file on the left. With considerable interest, not unmixed with anxiety, I saw a glittering line of steel extend through the thin woods and cover our front. I saw, I think, the first puff of powder smoke and a bullet hit the stump on which I sat. A large beech tree was opportunely near me, and I immediately sought the protection of its trunk. As the puffs of smoke increased, the beech tree seemed to wonderfully decrease in size. But for personal reasons, I stuck to it. Captain Albert G. Jenkins, afterwards Brigadier General, came up the line of skirmishers, with his hat off, and the blood streaming down his hair and neck, and called for some one to go and get his horse, tied to a stake behind Hale's Battery. He did not, like King Richard, promise a kingdom for his horse, but I was thinking of the kingdom to come, and a chance to dodge it. So I left the beech tree, and ran through the brush, over the hill and mounted the horse. I rode up to the battery and saw a dismounted cannon being propped up for service by a lot of determined men. I asked one of them, "Where is my brother?" "Who is your brother?" "Lieutenant Welch of this Battery."

"There he lies. He has done his duty." Then I looked where the soldier pointed, and saw my brother upon the ground lying where he fell, with his head almost severed by a flying piece of iron from the cannon that he was aiming when it was struck and dismounted by a cannon ball. As he lay with both arms extended in the form of a cross, he reminded me of Christ crucified. One died for all mankind, the other for his native state, with the same willingness.

I rode the horse to where I left Captain Jenkins, and when I tried to dismount, I could not get my foot loose from the stirrup, and he could not mount. I was very much afraid that the tangle would be undone by the bullits, but solved the riddle by pulling out my camp knife and cutting the stirrup leather in two. I then repaired rapidly to my friend, the beech tree, and Captain Jenkins went his way in the fight, while I got the stirrup off my foot.

About this time a lot of men pushed in our left with blue trimmings on their uniforms. One of them fired at me, and I yelled to my next file on the right, that we were being out-flanked by the Yankees. The man on my right taking the same view of the situation as I did, with a sudden aim, he shot one of the supposed Yankees through and through. I do not remember how the mistake was rectified, but it was, before the poor boy died. The artillery of the enemy for some time had been making the most noise, but suddenly we heard a new sound. It came from the "Peace Maker" (a gun cast by Mr. Job Thayer at his foundry in Malden). The new sound was caused by the miscellaneous missels it blew at the houses across the creek, behind which the enemy was fighting. Trace chains, mashed horse shoes and other kinds of scrap iron made the boards and shingles fly—and the Yankees also.

About sundown, the order passed to the skirmishers to rally on the center, which we did, and Lieutenant Nicholas Fitzhugh led us across the Green, and while we were burning some buildings to prevent their giving shelter again to the enemy, in case they should return, a lot of Federal officers rode up, supposing that we belonged to their army, on account of our incendiary occupation. One of them asked us where the d-------d Rebels were. We closed in around the lot, and gave them the information sought. They were Colonel DeVilliers, Col. Neff and others. They consented after some parley to visit Richmond. I then got permission from Lieutenant Fitzhugh to go across the hills to Upper Falls of Coal River, where my mother was at the time, and tell her the sad tidings of the death of her son.

Here ended my first lesson in the catachysm that followed.

Very Respectfully,
Levi Welch

The March Over the Mountain

Victoria Hansford was 22 years old when the Civil War started in 1861. She lived at Coalsmouth (St. Albans) near Camp Tompkins when the Confederate militia companies began to assemble there. As she got caught up in the excitement of the time she felt moved to record what was happening in her diary.

Since she lived so near the military camp she observed and wrote, among other things, about some early troop movements that have never been fully documented in any published Civil War records.

For instance, she wrote about one operation that took place nearly a month before the Battle of Scary:

MEN OF KANAWHA

That Government of Virginia which was destroyed at Richmond, has been reconstructed at Wheeling, and so acknowledged by the Government and People of the United States. At the call of Governor PIERPONT the President has sent armies of our friends from Kentucky, Ohio, and Indiana and expelled our invaders from the East headed by a lawless ruffian who in his retreat has left everywhere marks of bloodshed, fire, violence, and pillage, carrying with him several unoffending citizens as captives. You contributed nothing to his expulsion. You are called upon now by every consideration of defence, loyalty, and patriotism, to rally. Our Civil and Military Departments must be re-organized. I have been commissioned by the Governor to muster volunteer companies into the service of the State. You must also form yourselves into companies for service under the United States. Turn out, turn out, and range yourselves under one or the other banner, and never let it be said we were made free and cannot keep ourselves so. As fast as companies are formed they will be armed, and presently we shall be able to say to our liberators, go with our blessing, to serve your country else-where, we can defend ourselves.

L. RUFFNER, Adjutant.
Charleston, Augst, 1861.

One morning about the last of June I heard loud cheering at the mouth of Coal River. I threw on my bonnet and ran as fast as I could to see what was going on. There were four companies commencing to ascend Rust Hill across from the mouth of Coal. They went Indian file up the winding trail and across the ridge. Here and there they could be seen through the openings in the trees, each company showing different uniforms. First, Captain Patton's Kanawha Riflemen with their gray jackets. Then came the blue uniforms of Captain Gus Bailey's Fayette Riflemen. Next came Captain Andrew Barbee's Putnam Border Riflemen and bringing up the rear was Captain Charles Lewis' Kanawha Rangers. They all joined in singing "Virginia Boys" to the tune of Dixie. It sounded beautiful as the light breeze bore it down the mountain and across the river. But it brought tears to some of our eyes when they sang "We Will Die for Old Virginia."

☞ Secession sympathizers say the rebel loss at Scary was only four killed and thirty wounded. We learn from a reliable source that the rebel loss was not far from seventy-five killed and one hundred and fifty wounded, among them Capt. Patton, who had his right arm shattered, the same ball passing through the muscles of the right breast, creating a terrible wound, from which he is not likely to recover. On calling the roll after the battle four hundred of the rebels were among the missing. Albert G. Jenkins had his scalp lock grazed with a musket ball, but escaped uninjured. A rebel soldier who was in the battle states that at least one hundred shots were made at the brave Capt. ALLEN before he was struck. Capt Barbee claims the honor (?) of having shot this brave soldier, but he lies. The fatal shot was made by a rebel standing near the foot of the hill, who was himself killed a moment afterwards.

The Gallipolis Journal, Oct. 9, 1861.

Steamboats on the Great Kanawha

Capt. William F. Gregory was seeing the Great Kanawha River as he had never seen it before. As he stood in the pilothouse of the big sidewheel steamer *Moses McLellan* with her master, Capt. William Knight, the river seemed smaller than it ever had.

The *McLellan*, about 220 feet long, was the largest boat ever to run on the Great Kanawha. She had departed Gallipolis, Ohio, loaded with over a thousand tons of supplies for the Federal troops encamped at Camp Piatt, 10 miles above Charleston.

In 1861, it wasn't just his steamer's size and the conditions of the river that concerned Captain Knight, but it was the war that now raged in the valley and the nation as well.

After arriving at Camp Piatt and quickly unloading his cargo, Captain Knight ordered Gregory to turn the boat downstream and return to Gallipolis. However, turning the *McLellan* around in this narrow stretch of river proved to be impossible. Captain Gregory began to back the big steamer downstream. After backing about 12 miles, Gregory finally found a section of river wide enough to turn the *McLellan* and proceeded on to Gallipolis.

At the outbreak of the Civil War, the Kanawha River became an area of interest for both the Union and Confederates. The James River-Kanawha River route to the Ohio River valley and its strategic importance was not overlooked by either side. In addition, the vast salt works represented quite a prize for both armies.

Five days after the Civil War had started, Virginia seceded from the Union and began to move troops into the Kanawha Valley.

Confederate Brig. Gen. Henry A. Wise and a small force of men consisting primarily of the Richmond Light Infantry Blues, the Pig Run Invincibles and two cavalry units arrived at Charleston on June 26, 1861, to secure the valley for the Confederate States.

In the meantime, Federal Brig. Gen. Jacob D. Cox was massing troops at Gallipolis, Ohio, to drive all Rebel troops and supporters out of the valley. In July, he deployed some of his troops along an overland route up the river while the main body of his troops were moving upstream aboard a flotilla of steamboats. This flotilla was commanded by Capt. John McLure. As a result of the success of this operation, Captain McLure was later promoted to Commodore of the Kanawha U.S. Fleet.

Travel on the Kanawha during late June and early July could be very slow and as the flotilla proceeded,

the steamboats would sometimes have to land, allow the troops to unload, and continue with caution to get over some of the shoals and bars. After clearing the obstacle, the troops would re-board and the boat would resume their journey.

After a small skirmish at the mouth of the Pocatalico River, the Union forces were engaged by a concentration of Confederates on July 17 at the mouth of Scary Creek just below St. Albans. Although this encounter lasted all day neither side won a decisive victory. General Wise, however, claimed "a glorious repulse of the enemy, if not a decided victory." After this action, Wise began moving his troops up the Kanawha and out of Charleston toward Gauley Bridge fearing entrapment.

In late July during this Confederate withdrawal, the small steamboat *Julia Maffitt*, with a flatboat in tow, was loaded with supplies and about 700 Rebel soldiers, many of whom were the Kanawha Riflemen, for movement up the river. The *Maffitt* had been built at Cincinnati in 1860 for the Charleston-Cannelton trade and was owned and operated by Southern sympathizers. Carrying the Rebel troops and supplies was more than just 'another' job for the *Maffitt*'s crew.

As the boat was steaming upstream, Union troops were observed moving toward the river down the north bank. (This would be just about where Dunbar is located today.) Turning downstream to escape the enemy, the *Maffitt*'s lookout reported sighting another Union force moving up the valley. At this point, pilot Phil Doddridge turned the boat's bow toward the south shore, rang the engine room for slow ahead and buried the *Maffitt's* nose in the muddy bank.

As the Confederate troops scrambled to safety under fire from the Union soldiers the following action occurred as reported in the *Cincinnati Commercial*, under the heading of *Further news from the Kanawha Expedition:* ". . . and Captain Carter, of the Cleveland Artillery, fired one shot into the Rebel steamer *Julia Maffitt*, which caused her boilers to explode, and she burned to the water's edge."

The end for the *Julia Maffitt* had come at the hands of the enemy but just three months earlier, the *Kanawha Valley No. 2* had been lost to Confederate forces in confusion that often occurs in war.

Captain S. C. Farley was hired to move the baggage of the 22nd Virginia Infantry from Charleston upstream to Cannelton, then head of navigation, on his steamer *Kanawha Valley No. 2*. As the steamer passed the men of the 22nd marching along the riverbank toward Cannelton, the boat was hailed but Captain Farley refused to land and continued on upstream. In response to his failure to stop, some of the troops fired at the boat, riddling it with holes and killing one of the passengers. When the *Kanawha Valley No. 2* landed at Cannelton, and the baggage was being unloaded, there was a commotion about the incident. General Wise ordered her taken across the river and burned at what is now Montgomery. For years after the war, the charred hull could be observed during periods of low water.

Throughout the war, steamboats were used by both sides as the valley was controlled by first one side

"*Our first day's sail was thirteen miles up the river, and it was the very romance of campaigning. I took my station on top of the pilothouse of the leading boat, so that I might see over the banks of the stream and across the bottomlands which bounded the valley. The afternoon was a lovely one. Summer clouds lazily drifted across the sky, the boats were dressed in their colors, and swarmed with men as a hive with bees. The bands played national tunes, and as we passed the houses of Union citizens, the inmates would wave their handkerchiefs to us and were answered by cheers from the troops. The scenery was picturesque, the gently winding river making beautiful reaches that open new scenes upon us at every turn.*"

GEN. JACOB D. COX
"Military Reminiscences of the Civil War," excerpts were in
Battles and Leaders of the Civil War.

Built in 1855 at Cincinnati, the steamer *Moses McLellan* ran in the Cincinnati-Memphis trade before the war. Sold to the Upper Mississippi River in 1862 to run as a LaCrosse & St. Paul Railroad Packet. She was rebuilt and named *City of St. Paul* in 1866. WISCONSIN STATE HISTORICAL SOCIETY

When the *Silver Lake No. 2* came to Charleston in May 1863, *The Kanawha Republican* wrote: "There came up the Kanawha, on Saturday last to our place, a formidable looking craft, clad in iron mail. Her guns, we should think, would do admirable execution against an enemy, at a pretty long distance. Capt. Rodgers, his officers and men, by their gentlemanly department and fighting spirit, have won the highest regard of our people."

The *Silver Lake No. 2* was built at Wellsville, Ohio, in 1861 and sold to the U.S. Army. Her name was changed to *Marion* in 1865 and a year later she sank in Montana. PHC

and then the other. Some of the boats used were the *Undine, Florence, Reliance, Horizon* and *Izetta*. On Sept. 25, 1861, a war correspondent for the *Cincinnati Times* reports leaving Camp Enyart on the Kanawha for Cincinnati aboard the steamer *Capitola*. The article states that 1,000 Union troops were moving up the banks of the river to rid the area between Camp Enyart and Gauley Bridge of "Secesh" cavalry.

The *Capitola* had been built at Wheeling the year before and spent most of the Civil War chartered to the U.S. Army. After the war she ran in the Nashville-Louisville trade.

The use of steamboats on the Kanawha during the war depended on the need of the armies. This is vividly illustrated in the recollection of Captain W. F. Bahlmann of the 22nd Virginia Infantry, Confederate States of America.

After being wounded and captured, Captain Bahlmann was being moved down the Kanawha to the Union hospital at Gallipolis, Ohio. His account starts after reaching the Falls of the Kanawha: "Most of the party were put on a steamboat but through some oversight Captain Thompson and I were sent to Cannelton, now Montgomery. At Cannelton I found Steve Riggs. He invited us to the house of his father-in-law, Captain Farley, an old river man. We had with us two young Federals, wounded at Lewisburg. I hinted to Mr. Riggs to invite these two men in. After we reached the ladies, they dressed the wounds of

Thompson and of me and I told them to dress the wounds of the two Federals which they did. I was trying to make fair weather for them.

"We all four took dinner with the Farleys. Mr. Riggs gave me a $5 bill. That afternoon a steamboat came and carried us down to Charleston and we were placed in a hospital where we stayed 11 days. . . . From Charleston we were sent by boat to Gallipolis, Ohio, where there was a regular military hospital. We reached Gallipolis somewhat early in the morning. I walked from the river with the officer of the guard while the others rode in an ambulance."

As the war continued, other steamboats were to experience the perils of operating on the Kanawha.

On March 29, 1863, the steamer *Victor No. 2*, under the command of Captain Fred Ford of Gallipolis, Ohio, nearly added its name to the lists of casualties of the Kanawha. While proceeding downstream with U.S. Paymaster B. R. Cowen on board with a large amount of funds in his possession, the *Victor No. 2* was hailed by an individual who was apparently alone near Halls Landing. Captain Ford ordered the pilot to make the landing and pick up the passenger. Unknown to Captain Ford, laying in hiding was a force of Confederates who revealed themselves as the boat neared the shore and began firing. The *Victor No. 2* was immediately backed away from the shore while being pelted by lead balls. She was able to move beyond the range of the rifle fire but not before being thoroughly riddled. Captain Ford continued on to Point Pleasant, Va., where he informed the Union commander, Captain Carter, of the presence of rebel troops. Captain Carter did not see fit to prepare any defenses for the town and was taken by surprise the following day by Gen. Albert G. Jenkins.

As late as February 1864 travel on the Kanawha was not always safe for the Union soldiers. Capt. Charles Regnier of the steamboat *B. C. Levi*, had just finished loading his boat with lumber at Point Pleasant on Feb. 1, when he was ordered back to Gallipolis from where he had just come. Here he spent the night and most of the next day before returning to Point Pleasant to pick up Union Gen. Eliakim P. Scammon. Arriving about dark, the general wished to proceed up the Kanawha to Charleston and request Captain Regnier to make the run that night. Because of high winds and stormy conditions, Captain Regnier informed the general that it was too dark and stormy to go beyond Red House Shoals. The *Levi* would have to pass the shoals by a dug chute and this would be extremely hazardous at night without enough light to see the walls. The captain explained that he could leave at 1 or 2 o'clock and still reach Charleston by morning. This did not suit General Scammon and he ordered the trip to begin at once.

Against the advice of the captain and his assistant quartermaster, Capt. G. J. Stealey, the general insisted and the boat pulled out into the river at 7 o'clock and headed upstream. Although the night was dark and many of the landmarks hidden from the pilots' view, the boat arrived below the shoals at a little past 1:00 a.m. The captain then went to his cabin to retire where he found the general pacing back and forth and very disturbed by the delay. He was informed that the boat would continue upstream as soon as enough light permitted and that the pilot was to be called at 2:30. However, when the pilot got to the pilot-house he decided that it was still too dark to proceed safely.

At about this same time a unit of the 16th Virginia Cavalry, C.S.A., under the command of Maj. James H. Nounnan arrived at Winfield and saw the boat laying near the opposite shore. A small boat was procured and Lt. E. G. Vertegans with twelve men jammed into the boat, silently stole across the river and boarded the *B. C. Levi*. After overpowering the crew and passengers, the *Levi* was run across to the Winfield side of the river where it remained until late

the next morning. A large amount of the cargo of medical supplies and twenty horses were removed from the boat. She was then run down to the mouth of Hurricane Creek and landed about a half a mile below at Vintroux Landing. Here General Scammon, Capt. William G. Pinckard, Lts. Frank Millward and William C. Lyons along with twenty-five noncommissioned officers and men were taken ashore. The crew and other passengers were ordered off the boat and the rebels put her to the torch with a loss of an estimated $100,000 worth of medicines and munitions. All of the Union privates and noncommissioned officers except one were paroled by their captors. General Scammon, Captain Pinckard, Lieutenants Millward and Lyons and Sgt. Thomas McCormick were taken to Logan Court House and then to Richmond.

The news of the capture was soon known and Union Gen. Alfred N. Duffie sent Colonel Hayes with one hundred men in pursuit of the rebels. The chase lasted five days but the rebels eluded the Federals at every turn.

While the flames of war burned their brightest, some steamboat owners tried to operate as often as conditions permitted. Among these were the *Allen Collier*, *Annie Laurie*, *Victor*, *Victor No. 3*, *T. J. Pickett* and *Kanawha Belle* (1st). Sometimes operating during this period had its risks. The *Victor* was attacked by a deputy U.S. Marshall in April 1862 for operating on the Kanawha without a U.S. license. While on Sept. 27, 1862, the *Allen Collier* along with the steamer *Belfast* went to Augusta, Ky., to act as gunboat when Morgan's Raiders invaded the town. However, upon learning that Morgan's troops had howitzers, the two steamers turned tail and left the scene.

With the end of hostilities, travel and commerce increased as the valley inhabitants interests returned to pre-war concerns.

The *T. J. Pickett* along with the *Market Boy* worked for the Union Coal and Oil Company of Cannelton, W.V., carrying crude oil in barrels from Cannelton to the refinery at Maysville during and after the Civil War.

In November 1865, the *T. J. Pickett* transported a group of New York investors interested in oil properties up the Kanawha to view possible investment sites.

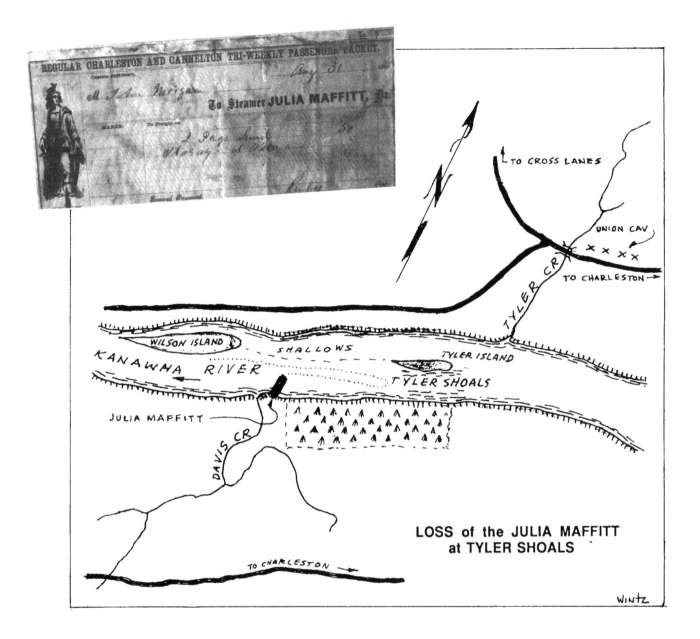

LOSS of the JULIA MAFFITT
at TYLER SHOALS

WINTZ

The Scuttle of the *Julia Maffitt*

After the Battle of Scary on July 17, 1861, the Confederate forces began their "Great Skedaddle" out of the Kanawha Valley. Union Gen. Jacob D. Cox was moving up the valley while the northern wing of his army was advancing from Ripley toward Kanawha Two-Mile at Charleston. At the same time, Gen. William S. Rosecrans was driving toward Gauley Bridge. Gen. Henry Wise, to avoid being cut off in the valley, pulled his Confederate forces all the way back to the heights of Gauley Mountain where he intended to make a stand.

The last Rebel troops to leave Coalsmouth consisted of about 700 soldiers of the 22nd Virginia Infantry. The only Confederate steamboat left on the river was the *Julia Maffitt*, which had been standing by at Coalsmouth. Soon after the Battle of Scary, the soldiers boarded the *Maffitt* after loading her with some of the left-behind supplies and equipment. They then managed to head up the river just before the armed Federal steamer *Economy* got underway at Poca. Cox's other troops had already moved out advancing toward Charleston by way of Cross Lanes.

The *Julia Maffitt* proceeded up the river until she reached the vicinity of Tyler Shoals. As she approached, the chute, which was near the south shore, she was challenged by Cox's advance troops who had already arrived from Cross Lanes. The Federals at first thought the *Maffitt* was one of their own boats and the Rebels could not identify the troops on land. However, when they began firing at the boat, the pilot, Phil Doddridge, quickly sized up the situation and took the only evasive action left for the *Julia Maffitt*. Backing her away from the chute in the shoals, he swung her around and headed for the low bank near the mouth of Davis Creek to get out of range of Federal small arms fire. He also knew the *Economy* was coming up the river not far behind and that she had an artillery piece mounted on her for-

ward deck.

What Doddridge didn't know was the Federals, on high ground above Tyler Shoals, also had artillery with them. As he headed for the south shore, they opened fire on his sturdy little boat and one round pierced the hull but did little damage. Reaching the Spring Hill side of the river, Phil Doddridge ran the forward deck up on a low section of the bank so the troops could get off in a hurry. As they scrambled up the bank, they found they had landed near a wheat field full of dry-cut wheat ricked in shocks. Evidently an officer directed some of the men to carry part of the dry wheat back on the boat where they scattered it around and set it on fire. A cargo valued at $100,000 and the boat were burned only minutes before a sizeable Union force landed on the scene. By then, Confederate troops that had been aboard the *Maffitt* were safely on the road to Charleston.

The *Julia Maffitt* burned to the water's edge and the hull eventually sank in place. Old-timers who lived near the mouth of Davis Creek claimed they had often seen the outline of her hull on the bottom before the stage of the river was raised.

Interstate traffic roars by 24 hours a day less than a 100 yards from where the bones of the *Julia Maffit* probably lies buried in the mud of the mouth of Davis Creek. This 1995 view from across the Kanawha shows the interstate Dunbar-South Charleston bridge. To the mid-left in the trees is the mouth of Davis Creek.

Original drawing by Dalton Moore.

Headquarters 3d Division, Dep't W. Va.,

CHARLESTON, WEST VA., APRIL 29th, 1864.

GENERAL ORDERS.)

No. 15.

I. Before a Military Commission, which convened at Charleston, West Va., on the 15th day of April, 1864, pursuant to Special Order No. 71, from these Headquarters, and of which Captain W. H. ZIMMERMAN, 23d Regt. Ohio Vol. Infantry, is President, were arraigned and tried:

1. Captain Geo W. Cox, of the steamer Victress:

CHARGE—"*Conduct highly prejudicial to the interest of the Government, endangering thereby the lives of its officers and soldiers, and rendering it's property liable to capture by the enemy.*"

Specification—In this, that Captain Geo. W. Cox, of the steamer Victress, a vessel then in service of the United States, did, on or about the night of the eighteenth of March, 1864, land and tie up the said steamer, some two miles and a half above Red House, a dangerous point on the Kanawha, and leaving the vessel at the mercy of the enemy, crossed over the river to Winfield, on a visit to his wife, and this was done in the face of the recent capture of Gen. Scammon and other officers, and the burning of the steamer B. C. Levi, and notwithstanding the best efforts of officers on board to prevent delay, and when ordered to put the boat, said steamer Victress, under way, the acting Captain refused to obey, saying that Captain Cox, on leaving the vessel, had ordered that she should not set out before four o'clock in the morning. The night was fine and clear, and the wind entirely lulled after ten o'clock.

To which charges and specifications the accused pleaded "Not Guilty."

FINDING—The Commission having maturely considered the evidence adduced, find the accused, Geo. W. Cox, Captain of the steamer Victress, as follows:

Of the Specification, (except in leaving the boat, and failing to run the schute after the wind lulled,) Not Guilty.
Of the Charge, Not Guilty.

SENTENCE.

And the Commission do therefore sentence him, the said Captain Geo. W. Cox, of the steamer Victress, "That he be not employed again in the Government service, and that he be not allowed to run a steamboat on the Kanawha river.

The evidence in this case does not warrant the finding. The whole proceedings are disapproved, and Captain Geo. W. Cox is released from his bonds.

Capture of Steamboat *Levi* and Gen. E. P. Scammon at Winfield

This incident occurred when the government steamboat *Levi* was captured and burned by Confederate troops and the capture of Gen. E. P. Scammon. The two articles are from official records of the Civil War, as follows:

REPORT OF THE CAPTAIN OF THE CAPTURED STEAMBOAT *LEVI*

I was ordered from Gallipolis to Point Pleasant, Virginia, on February 2, 1864, where I took on board General E. P. Scammon, who insisted that we leave that night for Charleston. I told him it was too dark and stormy to run farther than Red House Shoals, there being a dug shute that we could not run without it being light enough to see the wall. I told him we could leave at one or two o'clock in the morning, but he thought it best to go, so we left about seven o'clock, ran to Red House and tied up at one o'clock. The pilot was called at two-thirty o'clock, but said it was too dark and he could not see. While waiting for it to clear the boat was captured. A lieutenant and thirteen men came on board with a rush and secured all arms that were in the boat. There were some soldiers aboard, to the best of my knowledge about sixteen or seventeen. There was no sentinel on shore. At the time of capture of the boat, I had steam up, pilot at wheel, all ready to start.

After some little time, with a guard over the pilot and engineer the boat was run over to the Winfield side, and remained there until about ten o'clock and more of the enemy came on board making them number twenty-eight. We were taken some minutes after five a.m.

About ten o'clock the boat was ordered to run down to the mouth of Hurricane Creek. Soon after 12 o'clock the general and officers were taken ashore, and notice given to the crew and all to get ashore, as they were going to burn the boat. It was burned at 12:30 p.m. At that time is the last I saw of the general.

C. Regnier,
Captain, Steamer *Levi*

Gen. E. P. Scammon.
PHC

CAPTURE OF GENERAL SCAMMON, AT RED HOUSE VIRGINIA FEBRUARY 3, 1864

Reports Major James H. Nounnan
Logan Court House, Virginia
February 7, 1864

General:

I left Colonel Ferguson in Wayne County on the 26th ultimo without definite orders and moved in the direction of the Kanawha River along which I maneuvered in the counties of Putnam and Mason until the third instant when I entered Putnam court house at three o'clock a.m. with forty men and found a number of government officials and a government steamer, the *Levi*, with a strong guard and pieces of artillery lying upon the opposite side of the river. With great difficulty I secured a small craft capable of carrying four men only, with which I crossed a party of 12 men under Lieut. E. G. Vetregans, who obeyed my instruction as speedily as possible by cutting the telegraph line and assaulting the boat; which surrendered without firing a gun, although mounted several yards from the shore.

I found a valuable cargo on the boat, consisting chiefly of medical stores; a lot of arms, etc., and Brigadier General E. P. Scammon, Capt. W. G. Pinkard, and Lieut. Frank Millwood and his staff, and Lieut. Wm. C. Lyons, 23rd Ohio Volunteers and 25 non-commissioned officers and privates, who were made prisoners. I secured about 20 horses and some of the most valuable medicines, demolished the piece of artillery and a quantity of ammunition, paroled the non-commissioned officers and privates, except one and determined to make my way through with the most valuable prisoners with as much speed as possible. The enemy pressed me heavily in an effort to recover the prisoners and compelled me to come to this point.

I herewith send you, Gen. Scammon, Capt. Pinkard and Lieuts. Millwood and Lyons and Sgt. Thomas McCormick who refused to take a parole. I take great pleasure in bringing to your notice the coolness, discretion and courage displayed by Lt. Vetregans as well as good behavior and daring courage and fortitude of my entire force. I shall report to Colonel Ferguson of Wayne County as speedily as possible, and send the prisoners through from this point with a mounted guard.

I remain, General, with great respect your obedient servant,

James H. Nounnan, Commanding detachment 16th Va. Cavalry (CSA)

To:
Major General Samuel Jones
Department Western Virginia
Dublin.

The Great Kanawha Valley Flood

Since the beginning of recorded history many devastating floods have occurred on the Kanawha River. However, none have been as bad as the one that ravaged the valley Sept. 29, 1861. It could not have happened at a worse time for the already suffering people of the valley. It was the first year of the Civil War and both ill-equipped armies had just foraged up the valley commandeering all the food and livestock they could find. A great number of slaves in the valley had also just left their masters and many of them had not yet become situated with adequate food and shelter.

The rains started about the first of September in the mountains between Gauley and New Rivers and continued the entire month. When the flood crested at Gauley Bridge, the Union steamer "Glenwood" was able to go up over the Kanawha falls and deliver supplies to Federal troops four miles up the Gauley River. Some southern sympathizers began to lose heart after that as they believed it was a sign the Lord was on the Union side.

Union Gen. Jacob D. Cox had his headquarters in Charleston in 1861 when the flood occurred. In his book "Military Reminiscences" he commented on the big flood that crested there on September 29th at 16.0 feet above flood stage.

"The waters rose above these high banks, and flooded the town itself, being four or five feet deep in the first story of dwelling houses built in what was considered a neighborhood safe from floods. . . . There was enormous waste of supplies and loss of property but we managed to keep our men in rations, and were better off than the Confederates. . . ."

Solomon Minsker who ran a mill at Kanawha Court House (Charleston) wrote his brother in Cumberland County, Pennsylvania, giving him a first-hand report on the flood:

Kanawha C.H., Va.
Dec. 1, 1861

Mr. John Minsker
Dear Brother:

I believe I have not written to you since the big flood on the 29th of Sept. The Kanawha Valley was overflowed from hill to hill, the river was at least 15 ft. higher than ever I saw it. You know the drum house at Fields Creek, at the river, well the water was over the top of the coal cars as they stood on the railroad there. Every inch of ground in Charleston was under water and many, frame houses were washed away and some two-story houses were turned over on their sides. It done an immense sight of damage all along the river. I had 22 hogs drowned and at least two hundred bushels of wheat that was on the lower floor spoiled and about 40 BBLS of flour, but not all spoiled.

Down the river at Coalsmouth (St. Albans), Victoria Teays wrote how she experienced the great flood:

I was arroused about three o'clock by men in boats yelling, etc. got up and went to my upstairs window and I saw what I supposed to be a thick fog, but after listening and peering through the darkness I was afraid it was water. I threw a shawl around me and went downstairs to my father's room, who was then sleeping soundly. After gently awakening him I told him I believed the water was very high, even up in the road. He said I was foolish to think so and that I had better go back to bed. I sat a few minutes and then concluded to put on my shoes and go see for myself. Lo and behold, I found it some ten or fifteen feet inside our yard, the road was completely submerged long ago. I made haste back to the house, as it was very cold and chilly. Father could scarcely believe it when I told him and said it was unheard of, none of the oldest citizens had ever heard of its being out of its banks. He made haste to get into his pants and boots and took his cane and went to see for himself. He was amazed; stuck his cane at the edge of the water and said it would come no farther up in the yard. The next time he went to see about it his cane was beyond his reach.

And so it came up, up, up, and daylight revealed a fearful sight. The morning was cloudy, cold and foggy, there was so much water it was hard for us to get anywhere. Uncle Alva came up from his home just below the mouth of Coal River in a skiff and took me up the street in the boat where water had never been before. We landed across the street from Mr. Wheeler's store. All the families were moving out, the water was already in their houses. The river was already filled with all manner of stuff, farm produce, hay stacks, wheat, fodder, boats, houses, barns, chicken coops, corn cribs, etc. Such a sight I never saw before. The water rose steadily all day, but not quite so rapidly as it did in the morning.

We began making preparations for moving but this was almost impossible as we were totally surrounded by water. I had Jane, our cook, to get supper before the water came in the kitchen as it was even lower than the main house. The water from Tackets

Creek was coming up in back of the house. She had to put on her husband's boots to finish preparing the meal. We all then ate before the water could get into the dining room.

My brother Charley concluded to go out with the first passing boat and go up to our cousins', the Hansfords. Father begged me to do the same but I declined to do so unless he would go with me. He said the Yankees had never run him out of his bed and he would not let the water do it, so I stayed with him.

The colored folks had put their bedding and other things that the water might ruin upstairs in one room and I was in the other. About ten o'clock that night as father and I sat around the fireside, the water began to creep in around the hearthstone. Imagine our feeling as the water was from mountain to mountain across the valley with Kanawha, Coal and Tackets Creek all one great sheet of water. We had no boat in case of entire inundation, no one to help us.

We put some of the heavy pieces of furniture up on two chairs, but most of it had to stand in water. We took up the carpets when the water began to stream across the floors. Father got in his bed and I went upstairs to mine but there was little sleeping that night. We had taken two of our faithful dogs in the house when there was not one inch of dry land left. They became restless and began to prowl and howl, making the night hideous. There was one high point of land nearby where all the horses and cows had retreated. It was also covered but the water did not get over the animals heads so they were saved, but they kept up a dreadful noise all night. Several times during the night I called down to father through a closet, "How are you getting along?" and he would answer "I am still dry." However, the gurgling and washing of the water against the house was frightening. And so the night wore on and finally at daylight we heard the welcome sound of boats coming to our assistance.

Father waded out to the edge of the porch carrying his socks and boots to put on in the boat. The men waded in through the house to the foot of the stairs and called for me to come down and that one of them would carry me to the boat at the front door. As I would not allow any of them to carry me, old Uncle Jake Douglas arranged chairs in a row which let me walk out dry-shod. The water was very cold and we were chilled but we were taken clear across to Cousin James Teays, where we got our breakfast. Their house was filled to overflowing and Cousin Mary Ann Teays was quite cheerful and happy. She said she was glad to be able to return some of the kindness she had received when they had been compelled to refugee the past July when the advancing Union Army and the Battle of Scary had forced them to leave their home.

A party of us young folks got into a boat, for the water was calm by this time, and we went down to Aunt Thenie Wilson's. I got out of the boat onto their second-story porch. They were all upstairs, black and white, cooking and eating. It was even worse with them than it had been with us, as they lived exactly at the mouth of Coal River and the current was still strong there. The driftwood and logs had run through the lower windows, oh, it was a fearful sight. While we were there a large Union steamboat pulled in near the house as if to do all the damage it possibly could. From the disturbance and waves it stirred up, several outhouses broke loose and floated away. Imagine, the water from Rust's Mountain on one side of the valley to Teays Cemetery Hill on the other side, and we were nearly in the middle of it.

My father sent for me in another boat as he had decided that we would stay back in our house that night. He had the servants to wash it out as the water receded. They also built large fires in all the rooms and soon had it dried out. It had not been under water very long—from ten at night until about four o'clock the next day. So we had a good hot supper that evening and made ourselves very comfortable, considering all things.

There was a substantial old house still standing on the river bank just above the Amos Power Plant that was marked with the only known visible evidence of the great flood of 1861. The old home had continued to be in the Simms family since it was built in 1840. It is was the home of Mrs. Emma Simms Maginnis, a great-great-granddaughter of the original owner, who died a few years ago at over ninety years of age.

The old two-story frame house had a massive stone chimney on one end with a red mark painted on a corner up near an attic window. According to Mrs. Maginnis, the red line marked the crest of the great flood of 1861. She said it was put their by her family during the high water and it has been repainted many times since.

Mrs. Maginnis also remembered that in 1918 when the engineers were surveying for the big powder plant at Nitro they used the chimney mark to establish their high water reference. The official record at Nitro indicated that the high water elevation was set at 601.37. Since the elevation of First Avenue at 41st Street is approximately 585 feet above sea level, the flood of 1861 would have been about 16 feet deep at that location.

John Morgan, the father of venerable Sid Morgan, wrote a book "The Last Dollar" in which he briefly described his recollections of the 1861 flood. The Morgan farm was located where the Amos Power Plant now stands. His reference to the flood was as follows:

"The unprecedented flood of '61, which was ten feet higher than ever before known, swept nearly everything off the farm. The fences, Negroes, horses, cows, hogs, and everything were gone. Never will I forget how bleak, dreary, and desolate it looked."

Finally, the book "Men Mountains and Rivers" by Leland R. Johnson and published by the Huntington District, U.S. Army Corps of Engineers, also referred to the great flood.

☞ The Ohio is brim full and in some places running over its banks, caused by the recent heavy rains, which appear to have been general. The Kanawha on Sunday last was six feet higher than was ever before known, and the sudden and unprecedented rise was productive of immense destruction of property. Houses, barns, fodder, bridges, water crafts, &c., were swept away, and the suffering at the head waters of the Kanawha must be very great. Camp Enyart, six miles above Charleston, was inundated, the water being at the top of the ridge poles of the tents, and we learn that quite a heavy stock of army clothing received a soaking.

In 1861 the Kanawha River had risen at the furious rate of four feet an hour, creating a current so strong that the Ohio River seemed to reverse its course and appeared to run upstream from Pt. Pleasant to Letart Falls. Floating buildings raced out of the mouth of the Kanawha, smashed into the north bank of the Ohio, leaving the wreckage to be carried upstream together with other debris to form a drift pile that completely closed the Ohio River channel for a time. In 1861 Charleston had only 1,200 inhabitants and a few mills, but losses to the flood were heavy. The Kanawha Salt industry lost its boats, wharves, and buildings, and this, in combination with the interruption brought about by the Civil War, ended the prosperity of the industry.

One country preacher in Putnam County wrote the following in his Farmer's Almanac: "And so the great flood finally receded as the Lord had promised when he said: 'Never again will the floods come and destroy all life. For I will see the rainbow in the cloud and remember my eternal promise to every living being on earth'."

The Great Kanawha River flood as reported in the Sept. 28, 1861, of *The Gallipolis Journal.*

On Dec. 20, 1861, the first military execution in western Virginia—and one of the first of the war—took place on Charleston's West Side. Pvt. Richard Gatewood of Company C, 1st Kentucky Infantry, was stationed at Gauley Bridge under Gen. Jacob Cox's command, when he was court-martialed for desertion, threatening an officer and assaulting a fellow soldier. General Cox personally took charge of the execution in the interest of maintaining discipline among his troops.

Gateway to the Valley
Gauley Bridge

The Bridges of the Gauley River

by TIM MCKINNEY

Six bridges have spanned the mouth of Gauley River and it is from those bridges that the community of Gauley Bridge drew its name. The first bridge here was constructed in 1821 and was destroyed by fire in 1826. In 1828 the second bridge was completed and served until 1850 when it was replaced by an impressive new covered bridge. This new bridge was a huge wooden affair resting on three stone piers. Its handhewn timber, oak shingles and weather-boarded sides made it by far the best bridge ever to span the Gauley up to that time. This of course was a toll bridge, one of the finest in western Virginia. Collector of tolls was J. H. Miller Sr., merchant and well-known businessman. This was the bridge that was in use when the Civil War broke out.

Known as the gateway to the Kanawha Valley, Gauley Bridge was occupied by Confederate forces in June 1861 and the area became a hotbed of military activity. The hamlet of Gauley Bridge consisted of a few homes, a country store, a tavern and a church. When Federal forces commanded by Gen. Jacob D. Cox invaded the Kanawha in July 1861, retreating Confederate's commanded by former Virginia Gov. Henry A. Wise set the Gauley Bridge on fire. Its destruction was described by Addison B. Roler of the Wise Legion: "Several men were employed in tarring the side walls of the bridge and arranging loose material in it so as to burn with the greatest rapidity. . . . When all the companies were over the bridge with their baggage and commissary stores . . . the fire was

set at about 11 p.m. It burned very fast, the first arch that was fired fell in about one half hour. The whole length of the bridge was at least 150 yards, and ten minutes after the torch was first touched, the whole bridge was one sheet of flame . . ."

Two days after the covered bridge was destroyed Federal troops reached Gauley Bridge. On January 25, 1862, a new wire suspension bridge was completed over the Gauley River. This bridge was about 585 feet long, 10 feet wide, divided into three spans. The *Cincinnati Commercial* of February 17, 1862, carried the following notice concerning the new bridge: "The Gauley Bridge . . . has been rebuilt by Captain E. P. Fitch, the brigade quartermaster, attached to the staff of General Cox. It was constructed in 23 working days . . . and was opened for travel on the first day of this month. . . . The main sustaining parts are one and one quarter inch wire ropes. The roadway is of wood. . . . The entire work was executed by Stone, Quigley & Burton, bridge builders of Philadelphia."

To test the safety of this new bridge more than a thousand soldiers including infantry, cavalry and artillery, were marched and countermarched over its entire length. The regimental band of the 28th Ohio Volunteer Infantry joined their comrades on the bridge, everyone marching to the music of the band at a cadence step. Needless to say, the test proved satisfactory and the bridge was opened.

On Sept. 10, 1862, Confederate forces commanded by Gen. William Wing Loring attacked nearby Fayetteville, defeated the Yankees there and arrived opposite Gauley Bridge the following day. Retreating Union soldiers under command of Col. Joseph A. J. Lightburn cut the wire ropes of their recently completed bridge and watched as the entire work fell into the Gauley!

For six decades there would be no bridge across the Gauley. Finally, in 1926, a new bridge opened greatly enhancing travel along U.S. Route 60 the Midland Trail. This bridge was replaced in 1950. Its replacement was the sixth bridge to span the mouth of Gauley River and it is still in use today.

The first bridge over the Gauley River was built in 1850 and burned by Confederate troops in 1861. TMC

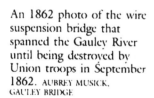

An 1862 photo of the wire suspension bridge that spanned the Gauley River until being destroyed by Union troops in September 1862. AUBREY MUSICK, GAULEY BRIDGE

One pier is still standing in the Gauley River from the two bridge built spanning the river in 1850 and 1862.

Another view of the wire suspension bridge across the Gauley River. WVSA

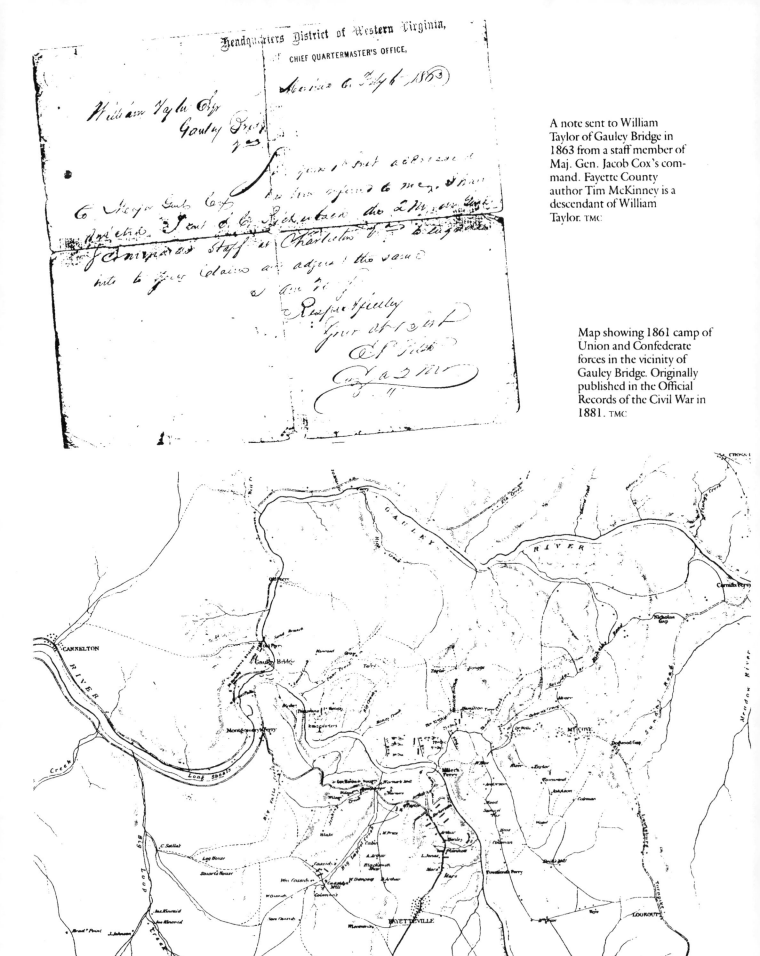

A note sent to William Taylor of Gauley Bridge in 1863 from a staff member of Maj. Gen. Jacob Cox's command. Fayette County author Tim McKinney is a descendant of William Taylor. TMC

Map showing 1861 camp of Union and Confederate forces in the vicinity of Gauley Bridge. Originally published in the Official Records of the Civil War in 1881. TMC

The Falls of the Kanawha near the confluence of the New and Gauley River at present-day Glen Ferris. This painting was reproduced in color in Edward Beyer's *Album of Virginia*, 1857.
VIRGINIA STATE LIBRARY

View of Kanawha Falls showing tents and wagon shop of the Union army. This view was taken from the south side of the Kanawha River.
AUBREY MUSICK, GAULEY BRIDGE

A modern view of Kanawha Falls, looking south. Little has changed in the past 130 years.

Report of Capt. Bailey
Scouting on the south side of Kanawha
Aug. 28, 1861

Report of Capt. Bailey, of his scouting between Miller's and Montgomery's Ferries, made to Gen. Wise, this 28th day of Aug. 1861.

On Monday the 26th instant. I went out on the cliffs of New River opposite the Hawk's Nest to examine the works of the enemy at that place and to ascertain their numbers. I did not discover any fortification or works for defense, but I saw two companies of soldiers. Both companies were infantry. I counted in the one company eighteen files and in the other seventeen, marching four abreast, which would make about one hundred and fifty men in the two companies. When I first saw them they were near the house of Daniel N. _____ They then formed and marched up the road until they passed from my sight, owing to a clump of trees intervening. I changed my position and in about one half hour afterwards I saw two companies of about the same size pass down in the direction of the mouth of Gauley River. I supposed them to be the same that I had seen going up the road.

I camped in the wood that night and went on the next day to Cotton Hill. I passed out on a ridge to the right of the Giles, Fayette, and Kanawha Turnpike, followed it on the top for about three miles, passing three pickets of the enemy, stationed on the turnpike to my left at intervals of about one mile, each picket numbering from twenty to thirty men. I followed the ridge to a point just opposite the mouth of Gauley River. There I saw the enemy's position camp, works for defense and men. I was not able to count all the tents at the mouth of the Gauley, they being too thick and clustering together looking like a town. Below that place two miles at the house of Aaron Stockton I counted some *thirty* tents. I saw a flag flying from a tent near Stockton's house. The space for about three quarters of mile between the house & Gauley & next to the house was hidden from my view, so that I cannot state whether any tents were there or not. I saw a great many soldiers moving about the tents. I saw wagons with two horses each, going down the Kanawha River Road, which I supposed to be empty.

I counted thirty and a good many had passed before I commenced counting.

Their fortifications commence on the river's edge, where the Bridge stood, extending from the River towards a house occupied by Sam Miller, then making an angle and running up and parallel with the river for about two hundred yards to a point beyond which I could not see from because of the formation of the ground and the house of John Hill. Above and immediately behind that fortification and _____ there from one quarter of a mile I saw another, the ground leading to which is exceedingly steep. The last fortification commands that section of the road passing by Col. Thompkins' house, which section is about two miles distant there from said fortification. There were no cannon that I saw at the lower works, and at the upper I could not see them, but men seemed to be placed there as if waiting an attack. I could not tell of what materials the lower fortifications were built, it being entirely covered with dirt and the top being very wide. The upper one looked as if large stakes in logs of wood had been driven perpendicularly in the ground on the side of the hill, and the earth filled in from behind the stakes I distinctly heard hammering but could not tell the direction.

I am of the opinion that cannon can be taken to the point on the ridge from which I made my observations. The turnpike leads to the top of Cotton Hill and over it to Montgomery's Ferry. The ridge above mentioned commences at the turnpike some two hundred yards below the top of the hill. Along this ridge it would be practicable to construct a road over which cannon could be taken, by cutting down the timber in some places and digging in the side of the ridge and around some two or three rocky points. The road to be constructed would be between two and three miles long; though from my being very much fatigued the distance along the ridge may have seemed longer than it really is. From the point where I made my observations the distance to the enemy camp I would take to be about three fourths of a mile.

R. A. Bailey

Capt. Robert Augustus ("Gus") Bailey was Captain of the Fayetteville Rifles, Company K, 22nd Virginia Infantry, C.S.A. He was killed in the battle at Droop Mountain.

Van Bibbers Rock, at the Camp Reynolds site. Carving on a rock was made on Nov. 28, 1862, by Jno Day Jr., a member of the 23rd Ohio Volunteer Infantry.

This rare photo of Federal soldiers was probably taken in the Gauley Bridge area, but their command and date are in question due to the unusual weapons they carried. WVSA

Gauley Bridge taken during
the war.
AUBREY MUSICK, GAULEY BRIDGE

CAMP OF 5TH VIRGINIA VOL. INFANTRY, U. S. A.

FALLS OF KANAWHA, WEST VIRGINIA,   1864.

OUR CHAPLAIN

Gives each of us a copy of this Engraving, to show our friends the way we sing and hold meetings in camp. He desires us to tell them to pray for us and him, that we may prove faithful to our country and our God, and not be found wanting in any day of temptation and trial.

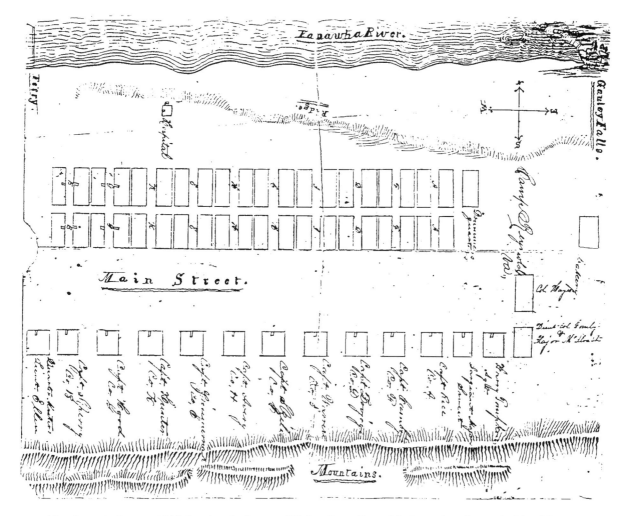

Map drawn in January 1863 showing the layout of Union Camp Reynolds, located on the south side of the Kanawha River at Kanawha Falls. DON MINDEMANN

Federal soldiers on Van Bibbers Rock, Camp Reynolds, Kanawha Falls, 1862. FAYETTE COUNTY HISTORICAL SOCIETY

WINTER QUARTERS.

Built by Col. R. B. Hayes in the Valley of the Kanawha, and occupied by himself and family in the winter of 1862–63.

FROM HOWE'S *HISTORY OF OHIO*

View of Camp Reynolds at Kanawha Falls, January 1863. RBHI

Homecoming
Battle of Charleston, 1862

The following is excerpted from "The Civil War Comes to Charleston," by Roy Bird Cook. This paper was read before the West Virginia Historical Society in annual session at Charleston on October 7, 1961. Dr. Cook passed away on November 21, 1961.

The Battle of Charleston

Colonel Joseph Andrew Jackson Lightburn marched his "army" into Charleston in April 1862, after an uneventful voyage from Ceredo, up the Ohio and the Kanawha.

Military operations in the Valley of the Kanawha around Charleston were now to proceed at an increased tempo. On August 6, Major Hall of the Fourth Infantry with a force of forty-eight men, had an engagement at Beech Creek, in Logan County, with a force of Confederates under Colonel Stratton and Major William Witcher. In the encounter Witcher was killed and several others wounded. Hall's body was returned to Charleston where he was buried but was later moved to Point Pleasant.

In the east larger military operations were shaping up which were to have a decided influence on the local situation. Events leading to the second battle of Bull Run, or Manassas, took place. The capture of a letter book of General Pope's gave the Confederate forces information of proposed Federal operations.The Confederate forces of Robert E. Lee invaded Maryland and threatened the national capital. General Jacob D. Cox, together with half his command in the Charleston-Kanawha region was ordered to move to the eastern theatre. The following order was issued:

Headquarters District of the Kanawha
Gauley Bridge, Va. August 17, 1862

General Orders No. 31
Brig. Gen. J. D. Cox hereby turns over to Col. J. A. J. Lightburn Fourth Virginia Volunteers, the command of this district * * *

Here by a stroke of a pen the tall, handsome, soldier-minister found himself in command of the Fourth and Ninth (West) Virginia Infantry, the Thirty-Fourth and Thirty-Seventh Ohio Vol. Infantry and two companies of the Second (West) Virginia Cavalry. His district included most of Southern West Virginia as well as the Kanawha Valley, his headquarters being at Charleston and Gauley Bridge.

The news of the withdrawal of part of the Federal troops soon came to the attention of General Lee who was informed that only about five thousand troops were left in the Kanawha Valley. On September 6, General W. W. Loring started on his march from Giles Court House (Pearisburg) toward the Kanawha. On September 8th General H. W. Halleck warned Lightburn of the impending invasion and noted that it might be necessary to fall back to Point Pleasant. On the 9th Lightburn wired Halleck suggesting that he be allowed to take up a defense position above Malden, east of Charleston, at the usual head of river navigation.

As events proved Lightburn did not have time to do anything but move "down the road." The Confederate forces moved in rapidly, having been preceded by the expedition under General Jenkins in the north central part of the present state of West Virginia. On September 10, Loring and his command attacked the

Federal advance post of Fayetteville and while an attempt was made to send assistance to Col. Siber, all outposts were withdrawn to the James River and Kanawha Turnpike and the Federal forces started down the Kanawha Valley.

The retreating Federals blew up the powder magazines at Kanawha Falls and also destroyed other military stores. A number of continual skirmishes took place between the incoming and outgoing forces. In the meantime Jenkins had followed the Staunton and Parkersburg road by Weston in a flanking movement designed to draw attention away from the Kanawha but to entrap the Federal forces. Jenkins finally reached the Ohio River at Ravenswood, crossed into Ohio, recrossed and by way of Ripley made passage over the Kanawha River at Buffalo, in order to harass the Federals from the south side of the river, as the army under Loring came down the river.

The entire military operations were brought on not only to draw attention away from the East but by a dire physical necessity—that reason being a very salty one. No salt could be secured to preserve meat in the Confederacy and the South was becoming desperate. Little wonder that attention was turned to the Kanawha regions where a short time before, 1,266,000 bushels of salt had been produced in a single year within a few miles of the towns of Charleston and Malden. General Loring commenting on the skirmish at Kanawha Falls, noted that "we took 700 barrels of salt."

The Federal troops moved down the Kanawha Valley in two divisions, one on each side of the river. Col. Siber arrived with the 37th Ohio at Brownstown (Marmet) on the morning of the 12th and here crossed the Kanawha to Camp Piatt (Belle), where his

Gen. William W. "Old Blizzards" Loring (1818–86) was Confederate commander in the 1862 Kanawha Valley campaign. He lost an arm in the Mexican War. Loring served under Robert E. Lee in the 1861 Cheat Mountain and Sewell Mountain campaign and under Stonewall Jackson at Romney in January 1862. After the war he trained the army of the Khedive of Egypt and never returned to his native land. PHC

Gen. Joseph Andrew Jackson Lightburn (1824–1901), a native of Lewis County and a boyhood friend of Stonewall Jackson, was the commander of Federal forces in the Kanawha Valley in the fall of 1862. After his flight from the valley in September 1862 he returned to Fayetteville in November, but was ordered to join General Grant for the Vicksburg campaign in December. He later took part in General Sherman's campaign in Georgia and ended the war stationed in northern West Virginia and Charleston. After the war he was elected to the state legislature and became an ordained minister in 1867, in which capacity he served until his death in 1901. PHC

Federal troops flee across the Kanawha River from the southside. (A composite scene assembled from Civil War drawings.)

troops joined the other wing directly under Lightburn. In the meantime the Second Brigade with Col. Samuel Gilbert of the 44th Ohio, moved down on the North side of the river arriving at Camp Piatt about 4 p.m. in a rainstorm. Four companies of the 4th Infantry were stationed at the mouth of Witchers Creek, with an outpost of cavalry. The operation served to protect the crossing of Siber's regiment and also to hold back the advance guard of the Confederates.

About 2 a.m. the morning of Saturday the 13th Gilbert's brigade moved through Charleston, crossed Elk River and took a position on the west side. Col. L. S. Elliott with a Federal detachment took position near the narrows just above Charleston but was later driven back by Col. John McCausland, with Col. Patton's 22nd Virginia and the advance of Lt. Col. Clarence Derrick with the 23rd Virginia Battalion. A spirited engagement taking place in what is now the area around the Capital of West Virginia. The infantry of the Confederates was strongly supported by Chapman's Battery of two pieces stationed on the side of the hill back of town. About 11:30 a.m. the Federals under Elliott withdrew to the center of town, their rear being protected by a battery of small Howitzers commanded by Lieut. Fischer of the 47th Ohio.

At the same time some Federal infantry fell back along the south side of the Kanawha being driven in by the advancing Confederates under Generals John Williams and John Echols. Lightburn who had taken up headquarters at the Joel Ruffner home place on present Kanawha Boulevard (1536), had in the meantime notified civilians to get out of town. Some sought refuge on Cox's Hill (Cemetery Hill), but soon found that they were under fire from both sides, as the advancing Confederates soon occupied the higher level.

In the rear of the Ruffner home stood a log barn near a rail fence. The Federals attempted to throw up some defense with the rails, moved up a smooth bore cannon and attempted to reply to the intense firing of a Parrott gun, and that of Otey's and Bryan's Batteries, stationed along the river bank just below the present Morris Harvey campus. One shell cut down a six-inch locust tree; another passed through the roof of the Ruby House and landed near present Morris and Quarrier Streets, damaging the Rand garden. It seems clear here that the Southern gunners soon realized they were firing on "Friends" and moved the batteries down the south bank of the river, but Confederate batteries from higher points were soon playing havoc with the business section of the town.

About 3 p.m. Capt. H. T. Stanton and three men secured a boat, crossed the Kanawha and captured the garrison flag. By this time the Federals were leaving town and General Loring in person, arrived in "downtown" Charleston.

The scene of the battle of Charleston from the Federal side after 3 p.m. shifted to the west side of Elk River. Lightburn had, in the meantime, set up as rapidly as possible his plan of defense. This in general involved the prevention of the crossing of Elk River by the Confederates and to provide opportunity for a wagon train of over 700 wagons to move toward

Federal troops retreat across the Elk River suspension bridge during the Battle of Charleston, Sept. 13, 1862. Known derisively as "Lightburn's Retreat," it was a delicious revenge for Col. George S. Patton and his Kanawha boys, since they had been forced to leave their home town in July 1861. This scene was assembled from Civil War era prints.

Suspension Bridge Across Elk.

A 1901 view of the Elk River suspension bridge.

CHARLESTON
Sept.13th 1862 .34th.37th.44th & 47th.O V I

The wire suspension bridge across the Elk River was built in 1852 to connect the east and west parts of Charleston. On their retreat to the Ohio River, Federal troops cut the cables on the east end of the bridge, dropping the planking into the river and delaying the Confederate advance. After the war the cables were respliced and new planking added. On Dec. 15, 1904, a cable snapped and the deck turned over, falling into the icy river. Two children and 11 horses drowned. The bridge was called the Lovell Street (now Washington Street) Bridge. PHC

Ripley, Ravenswood and Point Pleasant. In this operation an attempt was made to defend the west bank of the Elk River bank, west side just above the mouth, with breast-works of logs from a local mill being thrown up hurriedly. The 34th Ohio Vol. Infantry took a position along what is now lower Kanawha Boulevard, facing the Kanawha River. With the suspension bridge and the main "turnpike" as a center the 4th (West) Virginia Infantry and the 37th Ohio Vol. Infantry took up positions on both sides of the road along Elk River. In the center of present West Charleston was a barn where the companies from the 2nd (West) Virginia Cavalry took their position. The two smooth-bore guns from Col. Siber's battery of four mounted Howitzers took position on the Watts Hill. In the meantime the fighting in the main town continued.

Thomas H. Barton, later a physician, set down an account of the battle as he saw it:

On the morning of the same day (13) Surgeon Ackley met us at Brownstown, where he procured a small flat boat on which were placed our provisions and hospital supplies. He also brought with him a squad of hospital attendants to assist in taking our supplies to Charleston. The surgeon labored like a private soldier. The river was very shallow and for ten miles we had the laborious task of rowing and pushing the boat along. We reached Charleston about noon and six or seven of the hospital attendants were detailed to take the boat and cargo to Point Pleasant. Intense excitement prevailed in the city. The streets were thronged with people many of whom were preparing to follow our army or leave the town for they feared the battle of Charleston was about to be fought over their heads. All of the government property for which there was transportation was now placed on a train (of wagons) and about two in the afternoon started in advance for Ravenswood on the Ohio River. About one o'clock Col. Lightburn crossed Elk River and the torch was applied to the government buildings containing the stores that could not be moved. The Confederates opened the engagement from a battery on the hill south of Charleston, our battery replying *** soon after the first gun was fired smoke was seen about a mile down the Kanawha. That was the boat carrying supplies.

Col. Vance with the Fourth (West) Virginia located above present west Washington Street on the west side of Elk River endeavored to protect the remaining Federals until they could cross the river. After Col. August Parry with the 47th Ohio and Col. Siber with the 37th Ohio had crossed Elk bringing up the rear,

the cables on the west end of the suspension bridge were cut permitting the end of it to fall. In the meantime the Confederate artillery had reached the heights of present Fort Hill, which in 1863 became Fort Scammon. The staff of General J. S. Williams took up positions here.

The true story may never be known but in the next move between two and three p.m. the Confederates started firing "red hot" projectiles. Some passed within twelve feet of the 34th Ohio who began a retreat across the west side meadows to the "pike." One shot tore up a fence and another hit the barn mentioned before and set it on fire. The artillery on both sides kept up unusually fierce bombardment, considering the number of guns engaged, mingled of course with continual infantry firing. Artillery kept on firing until five o'clock and the infantry engagement kept up until darkness came. In the engagement the Confederates had eighteen killed and eighty-nine wounded; from Lightburn's command twenty-five were killed and ninety-five were wounded and 190 were missing. During the afternoon the town suffered terrific loss from fire. The Methodist Church, South, used as a quartermaster's store, the Branch Bank of Virginia, the widely known Kanawha House, Brooks and Whittakers stores, and the famous Mercer Academy all went up in flames. Also several warehouses and cavalry barns were destroyed.

The evening of the 13th found Lightburn and a huge supply train on their way to the Ohio River. He had moved his troops, kept up a continual skirmish for fifty miles, fought one battle and saved over a million dollars worth of supplies. He had faced a superior force and marched his men over a hundred miles. The broken army reached Ravenswood then moved to Point Pleasant by boat. Lightburn's Retreat is one of the high points of the war in the Kanawha Valley and is still studied by arm chair military strategists.

Arriving at Point Pleasant by the 17th, Lightburn wrote out his own version of the whole operation, a long detailed excellent report.

One would have expected General Loring and his Confederate troops to move on down the Great Kanawha in an endeavor to take over the position at Point Pleasant. Loring, however, did not press his advantage and was satisfied to consolidate his gains by occupying the town of Charleston. His local operations were of great interest to the civilian sympathizers on both sides.

Down at Coals Mouth, a young woman, Victoria Hansford Teays, took in the local happenings and later

From Fort Hill, Confederate battery blasts retreating Federals on Sept. 13. The Elk River suspension bridge is in the background. (A composite scene assembled from Civil War drawings.)

recorded it in a diary. On the morning of the 13th of September she wrote, "My father said I am sure I hear cannon." This was true as the sound carried easily twelve miles. A woman ran to the Teays home to tell them to come to the mouth of Coal River that everything from Charleston was "on retreat." "Such a sight I never saw before or expect to see again," Miss Teays continued, "the river as far as you could see up and down was full of boats of all kinds and when I say the river was full of boats I mean just what I say. A person could almost have crossed the river by jumping from one boat to another." This was Lightburn's retreat. On the north side of the river the road was black with wagons and citizens going down the river in front of the army. But being a "southern lady" she noted "that all this time the cannonading was going on it was music to us."

On the other hand, Sarah Frances Young, down the Winfield road was not quite so happy about the matter. On September 14th she wrote in her diary, "Oh such bad news. We heard the Rebels were coming into the Valley with such a superior force the Federals could not drive them back and that the Union families are moving down the river fast." The next day, September 15, she continued: "We heard today that the Rebels had possession of Charleston and Pa has moved down the river somewhere." "Pa" in this case was Captain John Valley Young, widely known Captain of Company G., 13th Regiment, (W.V.) Volunteers. He had been in command at the mouth of Coal River but joined the army movement down the river where he wrote on October 6th from Point Pleasant: "We got here without any difficulty by twenty-four hours march, the Rebels at our heels."

"Such a sight I never saw before or expect to see again. The river as far as you could see up and down was full of boats of all kinds and when I say the river was covered with boats I mean just what I say. A person could almost have crossed the river by jumping from one boat to another."

VICTORIA HANSFORD TEAYS
Coalsmouth

The occupation of Charleston was a homecoming to many of the men in the Confederate Army. Col. George Smith Patton, of the 22nd Virginia, again

Richard Q. Laidley was born in Wheeling, Va., between 1834 and 1836. He practiced law and medicine in Charleston prior to being commissioned as a first lieutenant in 1861 and a captain in 1862. He was married to Miss Whitteker during the 1862 occupation of Charleston. Laidley was wounded at the Battle of Droop Mountain. In 1864 he retired to the Invalid Corps. He returned to Charleston after the war where he was a druggist. He died in 1873 and is buried in Spring Hill Cemetery, Charleston.

McCausland attempts to ford the Elk River, Sept. 13. (A composite of Civil War drawings.)

joined his family for a short time. Col. John McCausland, of the 36th was near his old home near Buffalo. One of the gunners in Bryan's Battery was none other than the noted Milton Humphreys educated at Mercer Academy, but fate held him up at Gauley Bridge on the day of the battle. He it was, who later, first used indirect firing of artillery near Fayetteville. Captain R. Q. Laidley, of the 22nd Virginia was back home, as was Captain McFarland, brother of the town's leading banker. General Loring, on September 22nd, wrote a message to George Randolph, Secretary of War, for the Confederacy, that he was sending some flags captured in Charleston, Gauley Bridge and Fayetteville as war trophies. In the old Whitteker home, the Confederates found a flag that had been made by the ladies of Charleston, in 1861 for the famous Kanawha Riflemen. Otey, Bryan, and Stamp lined up their batteries and for the time being the guns were silenced.

"Oh such bad news. We heard the Rebels were coming into the valley with such a superior force the Federals could not drive them back and that the Union families are moving down the river fast."

SARAH FRANCES YOUNG
Winfield

The next move on the part of the Confederate command, seems to have been to seize the printing plant of the local newspaper, of which John Rundle was or had been the editor. This it seems clear was the plant of the *Kanawha Valley Star.* In a few hours a long proclamation had been printed as a broadside and posted in the courthouse and other public places not seared by the destruction of fire and war.

Signed by General Loring, the long statement affirmed that the Southern Army did not intend to punish those who remained at home as quiet citizens but promised to do otherwise with those who followed the Wheeling restored government.

In addition to the several proclamations and general orders which the citizens soon received, a daily paper also appeared—the first daily of record in Charleston. This proclamation as well as several General Orders were issued as broadsides and also printed in *The Guerilla.* It was known as *The Guerilla* and carried this motto: "Devoted to Southern Rights

The late George W. Scott, a noted Charleston druggist, was eight years old when the Battle of Charleston was fought and, in a Sept. 13, 1940, interview with the *Charleston Daily Mail* newspaper reporter, R. W. Jackson, talked about details of the fight. Mr. Scott said his family was eating breakfast when the battle started but "it didn't take us long to flee to the hills." He said almost all of the town's population climbed Cox's Hill (the hillside north of the city which looks south along Capitol Street) to escape the fighting.

However, troops took up positions on the same hill and fired over the heads of the defenseless civilians. Mr. Scott said that Confederate guns, located on the heights south of town (Fort Hill), "played havoc with the town side." One of the balls from a six-pound gun struck a tree on Cox's Hill and all the youngsters, including Mr. Scott, made a dive for the cannon ball, trying to collect a souvenir.

and Institutions." Evidently the first copy appeared on Saturday, Sept. 29th, and states that it is "published every afternoon by the Associated Printers." The printers must have been found among the Confederate forces. The price was ten cents. Little local news appeared, and a statement was made that due to interruption of the mail no news was available from the East. General Loring issued an order that all persons coming into town report to Major Thomas Smith, Provost Marshal. A meeting was called of all salt producers; it was hoped that an agreeable price could be arrived at to be paid for in Confederate money, of course. Local men were urged to join a proposed "Flying Battery" to serve under General Jenkins. One hundred teamsters were needed and they were promised good wages and rations. The quartermaster also needed blacksmiths and wheelwrights.

The last number of *The Guerilla* located discloses that Major L. Gounart had lost his watch at the Goshorn Hotel and a reward of $100.00 was offered for its return. David Goshorn inserted an advertisement for two weeks offering to buy ginseng, beeswax, and fur. This issue also carried General Loring's proclamation of September 14th. It is possible that this was printed in each issue, but enough copies have not been located to draw this conclusion.

Loring had some success at first in securing recruits and supplies. Trains of wagons loaded with salt from Snow Hill and Malden started back to southwest Virginia, over the Giles Turnpike. In his report, however, Loring recorded:

The march of near one hundred and fifty miles and the detailing of forces to guard captured stores in the rear caused such abatement and exhaustion of my command as to compell me to halt at Charleston. This place too, being the point of departure of many later roads, in any event is necessary to be held.

Some rumors were heard that the Federals were coming back. To Richmond went complaints from citizens not only as to the handling of matters by Loring, but also of General John Floyd's command in Logan and Boone Counties. General Jenkins ventured to a point near Buffalo where he contacted the Federals on September 27th.

One of General Loring's orders during his brief occupation of Charleston in September 1862.

GENERAL ORDER.

**HEAD QUARTERS,
DEPARTMENT OF WESTERN VIRGINIA,
Charleston, Va., Sept. 24, 1862.**

General Order, No.

The money issued by the Confederate Government is secure, and is receivable in payment of public dues, and convertible into 8 per cent. bonds. Citizens owe it to the country to receive it in trade; and it will therefore be regarded as good in payment for supplies purchased for the army.

Persons engaged in trade are invited to resume their business and open their stores.

By order of

MAJ. GEN. LORING.

To the People of West-ern Virginia.

The Army of the Confederate States has come among you to expel the enemy, to rescue the people from the despotism of the counterfeit State Government imposed on you by Northern bayonets, and to restore the country once more to its natural allegiance to the State. We fight for peace and the possession of our own territory. We do not intend to punish those who remain at home as quiet citizens in obedience to the laws of the land, and to all such clemency and amnesty are declared; but those who persist in adhering to the cause of the public enemy, and the pretended State Government he has erected at Wheeling, will be dealt with as their obstinate treachery deserves.

When the liberal policy of the Confederate Government shall be introduced and made known to the people, who have so long experienced the wanton misrule of the invader, the Commanding General expects the people heartily to sustain it not only as a duty, but as a deliverance from their taskmasters and usurpers. Indeed, he already recognizes in the cordial welcome which the people everywhere give to the Army, a happy indication of their attachment to their true and lawful Government.

Until the proper authorities shall order otherwise, and in the absence of municipal law and its customary ministers, Martial Law will be administered by the Army and the Provost Marshals. Private rights and property will be respected, violence will be repressed, and order promoted, and all the private property used by the Army will be paid for.

The Commanding General appeals to all good citizens to aid him in these objects, and to all able-bodied men to join his army to defend the sanctities of religion and virtue, home, territory, honor, and law, which are invaded and violated by an unscrupulous enemy, whom an indignant and united people are now about to chastise on his own soil.

The Government expects an immediate and enthusiastic response to this call. Your country has been reclaimed for you from the enemy by soldiers, many of whom are from distant parts of the State, and the Confederacy; and you will prove unworthy to possess so beautiful and fruitful a land, if you do not now rise to retain and defend it. The oaths which the invader imposed upon you are void. They are immoral attempts to restrain you from your duty to your State and Government. They do not exempt you from the obligation to support your Government and to serve in the Army; and if such persons are taken as prisoners of war, the Confederate Government guarantees to them the humane treatment of the usages of war.

By command of

MAJ. GEN. LORING.
H. FITZHUGH,
Chief of Staff.

A CONFEDERATE
OF 1862.

HEAD QUARTERS, DEPARTMENT OF WESTERN VIRGINIA,
CHARLESTON, VA., September 14, 1862.
General Order No.

The Commanding General congratulates the Army on the brilliant march from the Southwest to this place in one week, and on its successive victories over the enemy at Fayette C. H., Cotton Hill, and Charleston. It will be memorable in history, that overcoming the mountains and the enemy in one week, you have established the laws, and carried the flag of the country to the outer borders of the Confederacy. Instances of gallantry and patriotic devotion are too numerous to be specially designated at this time; but to Brigade Commanders, and their officers and men, the Commanding General makes grateful acknowledgment for services to which our brilliant success is due. The country will remember and reward you.

By command of

MAJ. GEN. LORING.
H. FITZHUGH,
Chief of Staff.

On September 24th Loring issued another "General Order" printed in the local printing office, which was as follows:

Headquarters
Department of Western Virginia
Charleston, Va. Sept. 24, 1862
General Order, No.

The money issued by the Confederate Government is secure, and is receivable in payment of public dues, and convertable into 8 per cent bonds. Citizens owe it to the country to receive it in trade; and it will therefore be regarded as good in payment for supplies purchased by the army.

Persons engaged in trade are invited to resume their business and open their stores.

By command of
Major Gen. Loring
H. Fitzhugh
Chief of Staff

No doubt some of the relatives of Henry Fitzhugh, the above mentioned Chief of Staff and member of a prominent local family, took some interest in such proclamations, but the citizens as a rule did not. The town as such was not accustomed to war and citizens were bitter about the existing conditions. Something of the situation can be gleaned from the letters and papers of J. C. McFarland, head of the Branch Bank of Virginia and owner of the Kanawha House. On December 2nd he wrote to a Mason Campbell, former newspaper man of Charleston, then living in Washington:

You are doubtless aware of the sad destruction of property here on the advent of the Southern forces on the 13th of Sept. . . . The Federals on their retreat set fire to their large commissary warehouse, the fire taking in its range the Bank of Virginia Building, the Kanawha House, William A. Whittekar's large store, the warehouse, Southern Methodist Church, Academy, etc., myself being by far the greatest sufferer. The walls of the Kanawha House present a ghastly appearance. In the former building but very little and in the latter not a particle of furniture was saved.

The battle raging in town, the men, women and children fled to the adjacent hills and finally above Coal Branch on Elk River. . . . The loss of my private library, my books of account, Niles Register complete, all my papers, records, precious memorials the gatherings of the last 49 years of my life—was most deeply felt. They were all put up and neatly arranged in the office in the bank. I saved most of the valuables of the bank as they were in the vault which was fireproof.

Writing to Indiana Hornbook of Wheeling, January 5, 1863, McFarland recounted that the "Federal troops set fire to their warehouses and the Bank and the Kanawha House and others in the center of town were burnt thereby. By the coming in of the Confederates two things were accomplished, viz the getting of several thousand barrels of salt and the marriage of two young girls in town." So romance and war moved hand in hand—then as now. On March 17, 1863, still writing of the troubled days, he remarked that not a fence was standing, there were 100 teams in the streets, sidewalks were ruined. The Branch Bank of Virginia and the Kanawha House had occupied an entire block, but a big wind had blown down the fire-ruined walls of the hotel and all was one mass of rubble. The Presbyterian Church alone stood intact, but McFarland complained that its lecture room was being used for a kitchen for teamsters and, "our village presents a most forlorn and desolate appearance." Later in the summer, he wrote Governor Boreman that, "During all this period the Kanawha Valley has been in the occupancy of alternating armed forces. In September 1862, at the period of Lightburn's retreat, the banking house with its furniture and contents except those within the vaults were destroyed by fire."

Gen. John Echols (1823–96) was born in Virginia, but lived and later practiced law in Union, Monroe County. He succeeded General Loring in October 1862 as commander of Confederate troops in the Kanawha Valley. He was commander at the Battle of Droop Mountain in 1863 and fought with General Early in the 1864 Shenandoah Valley campaign. After Lee's surrender he escorted President Jefferson Davis on his flight from Richmond. PHC

CONFEDERATE STATES HOSPITAL,
Charleston, W. Va., September 15, 1862.

SIR: It is with great pleasure that I report to you the sanitary condition of your army. After a toilsome march over mountain range and valley, a distance of 169 miles, we have no cases of essential fever, developed either in camp or hospital, and but one or two cases of rubeola and parotitis, occurring sporadically, during this march. We fought the Federal forces first at Fayetteville with the following casualties: 16 men killed upon the field (1 lieutenant and 1 corporal in that number), and 32 wounded, 4 of this number, I may say, mortally. One man killed in the skirmish at Cotton Hill and 3 wounded, 1 of this number mortally. No one hurt at Montgomery's Ferry, except from the accidental discharge of a gun while crossing the river, wounding 1 man. Six killed at Charleston and 8 slightly wounded, making in all 23 killed and 43 wounded.

I may here call your attention to the conduct of the medical staff, whose duties required their presence with their commands, placing them in most exposed positions and liable to casualties in common with the soldiers. Their conduct was marked by great gallantry and most indefatigable energy in the discharge of their professional duty.

It is but due to the corps that I should specially call your attention to the conduct of Surgeon [S. C.] Gleaves, Assistant Surgeon [O. N.] Austin, and particularly to the daring exploits of Surg. Joseph [F.] Watkins at the ferry, swimming the river and saving the ferry-boat, capturing also one stand of colors.

The enemy's loss at Fayetteville, in killed outright, was 65 that we know of; their wounded could not be correctly ascertained, but it is known that three barge-boats were shipped from Montgomery's Ferry and passed Charleston *en route* for the Ohio, and that four wagons, filled either with wounded or killed, were burned along the road from Fayetteville to this place, leaving exposed, in the most inhuman manner, portions of partially consumed bodies on the road. We could not ascertain the number killed and wounded in the different combats on the road. Judging from the most correct information, they could not have been less than 180 wounded in that action. Four were left dead in Charleston and 5 wounded. Their loss west of Elk River, opposite Charleston, where they met with heavy loss, could not be ascertained, as the bridge across the river was destroyed to prevent our crossing, thereby enabling them to carry off their dead and wounded. The capture of hospital and medical stores cannot fall short of $20,000.

Permit me, in conclusion, to congratulate you upon the success of your arms and the health and working condition of your army.

Very respectfully, your obedient servant,

JNO. A. HUNTER,
Surg., C. S. Army, Medical Director Dept. S. W. Va.

Maj. Gen. W. W. LORING.

—

Supplementary return of casualties in the Army of the Kanawha, September 6–16, 1862.

Command.	Killed.		Wounded.		
	Officers.	Enlisted men.	Officers.	Enlisted men.	Aggregate.
22d Virginia, Col. Geo. S. Patton		1	1	8	10
86th Virginia, Col. John McCausland	1		2	18	21
45th Virginia, Col. William H. Browne	1	1		14	16
60th Virginia		1		1	2
51st Virginia, Lieut. Col. A. Forsberg		3		16	19
63d Virginia, Col. J. J. McMahon		2		3	5
23d Virginia Battalion, Lieutenant Colonel Derrick		2		2	4
26th Virginia Battalion, Maj. A. M. Davis, 45th Virginia, commanding		2		8	10
30th Virginia Battalion		1			1
Otey's (Virginia) battery		3		11	14
Stamps' (Virginia) battery				5	5
Total	2	16	3	86	107

[JNO. A. HUNTER,
Surg., C. S. Army, Medical Director Dept. S. W. Va.]

DISTRIBUTION OF SALT IN RICHMOND.

The following advertisement in the Richmond Dispatch tells a gloomy tale in a very few words:

SALT. The Council of Richmond having deemed it prudent to require that recipients of salt under the arrangement made by said Council in reference to that indispensable article (salt), should produce a certificate, sworn to before some person authorized to administer an oath as to their residence, size of family, &c., the undersigned having learned that many persons, amongst whom were the widow of Confederate soldier, have been made to pay a fee of twenty-five cents to obtain ten cents worth of salt, (two pounds,) proposes to devote three and half hours each day, during the continuance of the Council's very humane interposition in behalf of its fellow towns people, to relieve all person who may necessitated to accept of the privilege afforded by the Council aforesaid, from this inhuman requisition.

L. A. FODUS,
Notary Public.

A good example of the need for salt by the south in 1862.

No. 13.

Report of Col. John McCausland, Thirty-sixth Virginia Infantry, commanding Fourth Brigade.

HDQRS. FOURTH BRIGADE, ARMY OF WESTERN VIRGINIA,
Charleston, W. Va., September 18, 1862.

In obedience to instructions, I have the honor to submit the following report of the action of the troops under my command at the battle of Charleston, W. Va.:

While the troops were encamped at Dickerson's farm, I was directed by General Loring to take command of General Echols' brigade (he being sick), the Thirty-sixth and Twenty-second Regiments, Otey's and Lowry's batteries, and the cavalry under Major Salyer. I at once ordered Major Salyer to pursue the enemy, and I found him near Charleston when I arrived. We passed Camp Piatt, the Salines, Maulden, and other places, but found no enemy. Upon my arrival at a point near Charleston, I discovered the enemy's skirmishers posted behind fences and behind a barricade they had erected near the river. I at once deployed Lieutenant-Colonel Derrick's battalion as skirmishers, and advanced them so that the left would sweep through the town and the right rest upon the hills beyond. I supported the right with Colonel [J. J.] McMahon's regiment, the center with Colonel Rodgers' [Ponge's] regiment, and the left with Colonel Patton's. The reserve consisted of the Thirty-sixth Regiment, Lowry's battery, a section of Otey's, and the cavalry. It was stationed in the road near the river. The whole line advanced, with occasional skirmishing, to the banks of the Elk River, and there found the enemy posted upon the opposite bank, with all communication with the opposite bank cut off. They had destroyed the bridge. I at once determined to concentrate the troops on the extreme right flank and attempt to cross at a ford about 2 miles above town. We moved in that direction under cover of our artillery, which was posted on a hill commanding the enemy's position and also other parts of the field. Upon the arrival at the ford, it was found impossible to cross with infantry and artillery. I ordered the cavalry to cross and move down the opposite shore, and then moved toward our extreme left, where we collected boats and were ready, when nightfall put an end to the conflict. Strong pickets and support for the batteries were left, and the troops sent back to the wagons to get rations, &c., and sleep.

The next day we crossed and came to their camp. General Echols was kind enough to send me his staff. Captain Oatlett rendered me great aid. Captains Poor, St. Clair, and Roche assisted me and were prompt in communicating my orders. The officers and men acted well.

I am, sir, your obedient servant,

JOHN McCAUSLAND,
Colonel.

Col. H. FITZHUGH,
Assistant Adjutant-General.

No. 9.

Report of Brig. Gen. John S. Williams, C. S. Army, commanding Second Brigade.

HEADQUARTERS SECOND BRIGADE,
September 18, 1862.

CAPTAIN:

At day-break we resumed the pursuit, and found that his force had crossed the river before day, at Camp Piatt. I brought all the artillery to the front, and kept up a galling fire upon his rear as he moved down the narrow plain on the opposite bank. As we approached Charleston I discovered masses of infantry crossing the river to the south side for the purpose of checking our advance. I immediately sent Lieutenant-Colonel Clarke with his battalion of sharpshooters, supported by the Forty-fifth Virginia Regiment, who gallantly drove the enemy back, some fleeing down the river, others recrossing it. The enemy by this time had nearly completed the evacuation of Charleston, and were preparing to give us battle on the opposite bank of Elk River, behind their wagons and hastily thrown-up breastworks. A height on the south bank of the Kanawha, just below the bank of Elk River, overlooked and commanded the enemy's entire position, but his artillery commanded the road to this height, and his sharpshooters lined the opposite bank of the Kanawha. I sent Clarke's battalion, with some companies of the Forty-fifth, to engage these sharpshooters, while the artillery, under Major King, dashed by at full gallop, and, with but small loss, obtained the desired height, and from six pieces opened upon the enemy's right flank a most destructive fire. A few effective rounds drove the enemy from his position, and his regiments and wagons began a disorderly retreat, and nothing was left but his artillery to contest the ground. At this moment the suspension bridge across Elk River fell. I now sent Captain Marye with the information which my position enabled me to gain, suggesting that the bridge had been destroyed, but that Elk River could be crossed on flat-boats and the enemy's cannon taken. You at once put me in command of four regiments on the north bank of the Kanawha, with instructions to cross Elk River and take the enemy's batteries. This was rendered unnecessary by the enemy withdrawing his pieces and following his retreating column with the whole of his artillery.

Colonel Wharton, while associated with me, behaved with his accustomed coolness and courage. Major King managed his artillery with great ability, and displayed that calm courage so necessary to an artillery officer. Captain Stanton, my adjutant-general, rendered important service, and accomplished a feat of gallantry which should be remembered. While the enemy still occupied one-half of Charleston, accompanied by Lieutenant Hackler, of the Forty-fifth, and 3 men of the same regiment, [he] crossed the river in a skiff, under a heavy fire, hauled down the garrison flag of the enemy, and returned, unhurt, with the trophy.

This hurried account embraces all that now occurs to me worth mentioning of the four days' march and fighting from Fayette Court-House to Charleston.

I have the honor to be, sir, your obedient servant,
JNO. S. WILLIAMS,
Brigadier-General, &c.

Capt. WILLIAM B. MYERS,
Assistant Adjutant-General, &c.

No. 10.

Report of Col. William H. Browne, Forty-fifth Virginia Infantry.

CAMP BLAN, NEAR CHARLESTON, W. VA.,
September 17, 1862

SIR: I engaged my regiment no more until I got to Charleston. There I occupied the hills on the south bank of the river and had some sharp fun dislodging the enemy's sharpshooters from the streets and the opposite banks of the river.

The officers and men of my regiment deserve praise. They marched without a murmur and fought gallantly. And to you, general, who led us to the conflict, we feel that we have done our duty. Your own noble daring had its influence in prompting us.

Most respectfully, your obedient servant,
WM. H. BROWNE,
Colonel Forty-fifth Virginia Regiment.

THE DRAFT.

BY JENKS.

'Twas upon one midnight dreary, while I wandered worn and weary, wandered homeward, pondering various matters o'er. Of the draft I was thinking. From the draught of drinking I was not shrinking, shrinking rather from scenes of gore. The draft, it kept my heart a sinking. I voted it a bore. It was that if nothing more.

My knees were fairly shaking, and my very heart was quaking, as I revolved the fatal subject oe'r and o'er. Although I was on the point of gaping, with an appetite tho drums were still a tapping, tapping harder than before, as if they meant by dint of rapping to fairly bore me with apprehensions sore.

I risked being called a mullet, or perhaps a cowardly pullet, but then a Minnie bullet would hurt a great deal more. This and more I sat divining, on the doorstep half reclining, with the hall lamp dimly shining, and I repressed a snore. To sleep was fast inclining, dreaming one thing o'er, how could I escape that drafting bore.

As I before to you was stating, how I sat there meditating and narrating and debating the horrors oe'r and o'er. Ne'er to me were wars inviting, e'en when foes the dust were biting. I was always opposed to fighting, fighting as I said before. I won't fight if I cant help it, I swore a solemn swore, I'll fight the rebels, never more.

I must have fallen dreaming, for my fertile brain was teeming with curious sights, I ne'er had seen before. I wouldn't be dissension sowing, but some men I'm sure of knowing, were by me then seen going, going to the Surgeon's door; men who were both strong and healthy in the happy days of yore, but now coughing sick and sore.

Some were bent almost double as if with weary years of trouble who were straight and hearty the week before. Others pale and illy shaven with aspect mean and craven and with locks once black as raven, now with white were silvered o'er. Not a man looking less than forty four, and most of 'em a great deal more.

The doctor grew more dizzy, he was kept so very busy, inspecting each hernia and old forgotten sore. Unless they took to lying, half of them them were dying to judge by their feeble sighing, sighing as their looks implore the mercy of the Surgeon as he looked them o'er and o'er and examined every pore.

The diseases pulmonary and chronic dysentery, fistula and consumption's hollow roar. The peculiar intoning a sort of half moaning and sometimes a groaning, groaning from the stomach's core, was heard from early morn till after four, to be ceaseless evermore.

The doctor he was flurried, annoyed, tired, hurried by the tales that constantly there did pour. His nails he was a biting then he commenced his writing certificates indicting, indicting by the score. Each applicant an affidavit swore, then he showed them to the door.

Aha! quoth I awaking myself a shaking, I know how to evade that blasted bore. The ceaseless drum was beating as I stood repeating the new idea of cheating Uncle Sam a little more to escape the drafting bore. I'm an invalid forever more.

Portion of the Civil War Diary of Mrs. Henrietta Fitzhugh Barr (Barre) 1862–63, Ravenswood, Va., published by Marietta College, Marietta, Ohio, 1963.

Monday, September 15th. Last night the advance guard of the retreating army entered town and all day long they have been coming and "still they come." The baggage train is said to be sixteen miles long. Such an army I suppose was never seen, composed of negro men, women and children, refugees of all ages, sex and condition. We have been feeding the troops all day and until late at night. There are some apprehensions of a battle at this place, if the Confederates really are in pursuit (which I begin to doubt). They are retreating across the river as fast as possible. The negroes and the ambulances with their sick and wounded are sent first. Before night the valuable darkies are over the line, hence bid adieu to "Dixie." The soldiers say they were fighting five days all in a very exultant manner of having laid waste the country. Say they have no doubt there will be a famine. They seem to be a reckless set of men. Seem to have no idea of the possibility of retaliation or retribution.

Tuesday, September 16th. We sat up last night. It seemed unsafe to retire to bed with such large numbers of these dreadful men so close around us. In consequence we are unfitted for our regular duties today. We breakfasted about 15 of the men. After we had cleared away the tea things, went up to Susan's to see the remainder of the army pass. An immense train, then the artillery, next to the infantry, the cavalry followed. I did not wait to see them. We are filled with such intense anxiety about our dear ones at C. and dear brothers who have probably been exposed to these deadly battles that I cannot give a connected account of anything. Three steamboats came up the river to help take the army away. Part of them with the artillery embarked on the boats. The rest crossed the river and marched down, they say, to Point Pleasant. We passed an uncommonly quiet peaceful evening after the evacuation. Keeny, Armstrong and a great many of our Union neighbors have skeddadled. Poor guilty creatures, they are afraid of their own shadows.

Saturday, 20th. A day of bustle and confusion, mother having decided to go over to C. in a buggy with F. H. to see the boys. Mat, Susan and Sarah are preparing to follow her tomorrow. A company of cavalry calling themselves the "ragged secesh" numbering about 75 or so rode in and out of town in very excellent order today. They behaved well and quietly, disturbing no one. I hear they are stationed back of Ripley.

Friday, 26th. Mrs. Hoyt returned from Ripley in company with a small part of our cavalry. The latter went over the river and very near capturing the Yankee mail rider. If they had been on horseback could have succeeded in getting him.

Sunday, 28th. The news alluded to yesterday turns out to be the genuine truth. Dan Frost came last night and with him a bodyguard of men commanded by Lytle. The latter went out immediately after ordering a supper at Susan's and stealing all the cold provisions, buckets and tin cups she had about the house, to bushwhack on the road between here and Ripley. Their object is to secure any of our men who may be coming this way. In this they will be defeated. Proper measures have been taken to prevent mischief. Dan left this evening for Ohio . . . afraid to spend the night here. Anne and I went up to Mrs. F's after tea to discuss the various events of the past day. A startling development in the family of Mr. Fleming is the occasion of much talk. The weather is charming but very dry.

Swan Creek, O., April 2, 1892

Colonel Vance, Gallipolis, O.

Dear Old Friend—As near as I can figure from my bad memory, it was the 10th of September, 1862, that the Fourth West Virginia left Gauley for Fayetteville. I was detailed in charge of thirty men to occupy a trench dug on the side of the mountain, thirty yards from the road, and near two miles from Gauley Bridge, on Gauley River (called a rifle-pit).

I never received any word from the colonel in command to come in, or to vacate the rifle-pit, or that the forces had left camp.

On the 11th day of September, I learned from a citizen that my command had gone to Fayetteville to relieve the Thirty-fourth Ohio, and that you had retreated from Fayetteville, and returned to the Kanawha River, and had gone down the river toward Charleston. I immediately detailed a man to go to camp and see, and bring me orders what to do. In half an hour afterward I got word from the same citizen that the forces had vacated Summerville and gone across country for Charleston, not by way of Gauley Bridge. My man returned from Gauley, brought me word that the forces had retreated toward Charleston and left every thing in camp. I then pulled out from the rifle-pit, came to camp Gauley, and gathered the company books and the valuables we might find, and moved off for Charleston. When we got as far as the ferry, we found the place or buildings fired; also our ammunition house was burning. I saw the rebel forces marching down on both sides of the river, and in front of me, and a rebel cavalry closely after. I ordered a charge on the mountain, and the boys were all willing to obey orders. We kept the mountain top, and traveled on double-quick until about 10 o'clock at night, when it began to rain and was so dark we were compelled to camp. We pulled out from there as soon as we could see, and moved on until we came to where some creek came into the Kanawha River. We then came to the river road, and when within thirty or forty yards from the road, we saw what we supposed to be two regiments, one of cavalry and one of infantry, passing in our front. We concealed ourselves in some pawpaw bushes until they passed by, when I turned for Elk River, camping at 12 o'clock at night of the 12th of September on the waters of Elk River. There we found two bushels of Irish potatoes. Roasted potatoes until 3 o'clock, when I roused the boys and made a start for Charleston. I found the regiment about 3 o'clock in the evening of the 13th of September below Elk under fire.

Respectfully,
C. B. Blake

Letter from Lt. C. B. Blake, Fourth West Virginia Infantry, to Col. John L. Vance, lieutenant colonel Fourth West Virginia Infantry.

A boy named Thomas E. Jeffries stood all day on Cox's Hill and witnessed the entire battle. Young Jeffries who lived to become a U. S. Engineer, later wrote about his memorable experiences during that fateful day of September 13, 1862.

Came the night before when a steady stream of wagons passed down Kanawha Street and crossed Elk River. I got up and saw a crowd of people going toward the hill. Residents had been warned that the town would probably be set on fire and shelled by the retreating Federals.

I followed the crowd to a place near Spring Hill Cemetery above Piedmont Road. This area was known as Cox's Hill.

Someone suggested there should be a white flag hoisted, and from some unknown quarter, a garment that ladies used to put on first in the morning and take off last at night, was produced and it was fastened near the top of a pole that a martin box was on.

We could hear some firing and Steele Hawkins, a boy named Andrew Parks, and I decided to investigate. We passed through the cemetery and went some distance down the face of the hill and sat down under a tree.

The Kanawha Hotel, Bank of Virginia, Brooks Store, Methodist Church, a large warehouse and the Mercer Academy, were all on fire.

The Federal forces had passed toward the west side but we could still see both armies all on Kanawha Street headed for the old suspension bridge on Lovell St. . . .

While we were watching . . . a squad of Confederate skirmishers suddenly appeared coming up the hill. I had on a blue flannel suit and a blue cap, and Hawkins also had on a blue coat and cap.

When we saw them we jumped up and they, thinking we were Federal soldiers, fired at us. Fortunately the bullets only cut the leaves over our heads. I lost all interest in things down in town and started up the hill on the double and never stopped until I got into a crowd of women behind the hill.

Later Jeffries wrote:

We saw several dead Confederates lying on the grass next door to us. Several Federals were killed in a field at what would now be the corner of Brooks and Washington Streets. One man was killed near the corner of Lee and Broad Streets. He was in a garden eating tomatoes.

The Kanawha riflemen camped across the street from where the Baptist Temple now stands. My two buddies, Steele Hawkins and W. F. Goshorn, and I

NOTE: The Baptist Temple at the date of this article stood on the northeast corner of Capitol and Washington streets.

"The first I saw of the Confederate army was a small gun pulled by a large mule. While we were watching the small gun, a squad of Confederate skirmishers suddenly appeared, coming up the hill. There were two or three of us boys wearing blue flannel suits and caps, and were mistaken for Federal troopers. The soldiers fired on us but missed us, but we started up the hill in high and never stopped until we were behind some women."

THOMAS E. JEFFRIES
eyewitness

were frequent visitors at their camp and at times took a meal with them which consisted of a big cup of bean soup, a hunk of bread baked in a dutch oven, some fried bacon, and plenty of sop. Sop was made by stirring flour into bacon grease. One day Billy Goshorn sat down in the sop and we were run out of camp.

Victoria Hansford also described the Battle of Charleston as she remembered it from her home in Coalsmouth (now St. Albans).

On the morning of September 13, my father said to me, I am sure I hear cannons: When I went out to listen, I was sure I also heard cannonading far off. I said, "Oh, they are coming, they were coming home, and I must set to work putting my house in order."

About 10 o'clock Aunt Lucy came hurrying up the road to tell us that Mrs. Thenie wanted us to come down to the mouth of Coal. She said you never saw such a sight coming down the Kanawha as everybody in Charleston is on retreat. I got down there in a few minutes with Father and the servants coming along behind.

Such a sight, the river as far as the eye could see was covered with boats of all kinds. There were flat boats, jerry boats, jolly boats, skiffs, and canoes. In the boats were all kinds of people and all kinds of things. . . . A person could almost cross the river by jumping from one boat to another. They were not soldiers but citizens who favored the North and thought it was wise to retreat from Charleston with the Union Army. I felt sorry for them and said to myself, poor souls, the Rebs would not have hurt them anyway.

Sarah Young also wrote about the Federal retreat from the valley:

Sept. 15th, 1862

We heard today that the Rebels had possession of Charleston and Pa had moved down the river somewhere. Oh, Lord, in Thy kind mercy be with him and his company. Save them from the savage Rebels.

The Rebel forces that attacked Charleston included both the 22nd and 36th Virginia Infantry Regiments, which meant that most of the Kanawha Valley boys were coming home. Victoria also wrote about the homecoming:

By evening the Rebel Cavalry reached Coalsmouth. The next morning at scarce daylight, my brother came home. Oh, how thankful we were to see him; it was the only time he got home during the war. We gave up to the enjoyment of the present, entertaining our soldier boys in our homes. We also made them clothes so they might be warm in the winter as we all knew they could not stay in the valley. There was not enough to feed them and there was danger that the Yankees would return and cut them off from the rest of the army.

And so they must go, all was excitement as they were all moving out. We had to tell our friends farewell, for how long we could not tell, but we were prepared as everyone seemed to know it was a time for falling back. They all left in good order and by nightfall not a Reb could be seen.

The Confederates stayed in town about six weeks when they chose not to make a stand against a regrouped and strengthened Federal army that was poised to move against them. On Oct. 8, Loring's forces quietly withdrew toward Lewisburg, leaving the Union troops to reenter Charleston without a fight. All through the remainder of the war, the town remained under Federal control.

To add a happy note to the war in the valley, Sarah Young's last entry in her "Little Journal" read as follows:

November 19, 1863

Em Stacies and I were at Coals Mouth on a visit when to our great joy the 8th Regt. Vols. came in. I have a particular feeling for the 8th since I have one dear friend in that regiment. Edgar has been down several times since his regiment returned. Our friendship has grown to devoted love. Is there one in the world as noble? I have promised to become his wife soon. Now our Father, I ask Thee to make me in every way worthy of him.

During the war Sarah Young married Major Edgar B. Blundon from Morgan County, Ohio. He became a Methodist minister in the Charleston area where he died in 1873. Sarah Young Blundon had four daughters. She lived in Charleston until her death December 14, 1920.

"The Yankee Army came down the river on the run. Their wagons, horses nose to tail gates, and the drivers whipping them up. The infantry outran the wagons. When a team stalled, the wagon was thrown out of the road and the retreat went on. The Yanks were badly scared and everybody along the turnpike got out of the way and took to the hills. The Rebels were right behind them, pouring the hot shot into 'em."

ROBERT BANNISTER
private C.S.A. Artillery

☞ From the best information, the rebel force on Kanawha does not exceed 8,500, all told, which includes one regiment of negroes. The Gauley and Elk bridges, were destroyed by the Federal troops, and five houses in Charleston consumed by the contending forces at long range. Our men hadn't time to roll much salt in the river, and as a consequence half a million bushels fell into rebel hands. We've nothing more to say, only the victory is not quite so great as at first reported.

The Gallipolis Journal, Sept. 18, 1862

It is interesting to note the reference to one regiment of negroes. No information is available on this statement.

- 101 -

THE GUERILLA.

DEVOTED TO SOUTHERN RIGHTS AND INSTITUTIONS.

Vol. 1.   CHARLESTON, VA., SEPTEMBER 29, 1862.   No. 2

THE GUERILLA,

IS PUBLISHED EVERY AFTERNOON
By the Associate Printers.

TERMS—TEN CENTS per copy, or FIFTY CENTS per week.

For the Guerilla:
LINES ON THE MARCH.

A soldier lay on the frozen ground,
With only a blanket tightened around
　　His weary and wasted frame ;
Down at his feet the fitful light
Of fading coals, in the freezing night,
　　Fell as a mockery on the sight.
　　A heatless, purple flame.

All day long with his heavy load,
Weary and sore, in the mountain road,
　　And over the desolate plain ;
All day long through the crusted mud,
Over the snow, and through the flood,
Marking his way with a track of blood,
　　He followed the winding train.

Nothing to eat at the bivouac,
But a frozen crust in his haversack,
　　The half of a comrade's store—
A crust that after
Some pampered spaniel might have passed,
Knowing that morsel to be the last
　　That lay at his master's door.

No other sound on his slumber fell,
Than the lonesome tread of the sentinel,
　　That equal, measured pace,
And the wind that came from the cracking pine,
And the dying oak, and the swinging vine,
In many a weary, weary line,
　　To the soldier's hollow face.

But the soldier slept, and dreams were bright
As the rosy glow of his bridal night,
　　With the angel on his breast ;
For he passed away from the wintry gloom,
To the pleasant light of a cheerful room,
Where a cat sat purring upon the loom,
　　And his weary heart was blest.

His children came—two blue-eyed girls,
With laughing lips, and sunny curls,
　　And cheeks of ruddy glow —
And the mother pale, but lovely now,
As when upon her virgin brow,
He proudly sealed his early vow,
　　In the summer, long ago.

But the *reveille* wild, in the morning gray,
Startled the beautiful vision away,
　　Like a frightened bird of the night ;
And it seemed to the soldier's misty brain
But the shrill *tattoo* that sounded again,
And he turned with a dull, uneasy pain,
　　To the camp fire's dying light.
CHARLESTON, VA., Sept. 26, 1862.

LATEST NORTHERN NEWS.

We clip the following accounts of the fight at Fayette C. H., on the 10th instant, and the exodus of the enemy from the Kanawha Valley, from the Cincinnati *Commercial* of the 19th :

THE LATE BATTLE AT FAYETTE C. H.

From all I can learn, the only regiments engaged in the battle of the 10th inst., were the 34th and 37th Ohio. The Thirty-fourth was raised by Col. Don Piatt, and will be best known as the Piatt Zouaves. It is now commanded by Col. Toland, an officer who abundantly proved his military skill and capacity in the late engagement. The Zouaves fought desperately, and displayed an amount of courage and determination worthy of veterans. They met the rebels outside the fortifications of Fayette, while less brave men would have remained inside, and their long list of killed and wounded tell how manfully they battled for the cause of liberty. The force contending against them numbered not less that 7,000, at least half of which attacked them especially, while the others attempted the capture of the earthworks. The conduct of the 37th is spoken of in terms of highest praise. They led the enemy at different points within the fortifications, drawing them into cross-fires and ambuscades, and reducing their ranks terribly each time.

The loss of the 37th was small, and I have as yet been unable to obtain a list of those of its members who were injured. I sent by telegraph to-day the names of all the wounded of the 34th. The list of killed has not yet been made out ; it will number about fifteen perhaps more. Of the wounded very few are considered dangerous. A majority of the wounds are in the lower limbs. All the necessary amputations have already been made, and were noted in my dispatch yesterday. Capt. Hatfield, of company A, it is feared will not recover. He was shot through the hip, receiving a wound very much similar to that received by Gen. Nelson, at Richmond, Ky. The surgeon said to-day that the Captain's case was more hopeful than at first ; that he had not lost any thing in three days, and might possibly live. His wife and several friends are here to nurse and care for him. He is highly esteemed as an officer by superiors and inferiors.

Col. Toland escaped uninjured. He was at the head of his regiment during the battle, and had two horses shot under him.

The wounded Zouaves are well cared for. They are all in the general hospital, about a mile from the city—a building put up expressly for the purpose for which it is used, and admirably adapted to it. I have never seen a cleaner hospital or one where the wants of the sick are better attended to.

EXODUS FROM THE KANAWHA.

During the past few days the Kanawha and Ohio rivers, between this point and Gauley, have been full of flatboats, batteaux, skiffs, rafts, and all manner of buoyant conveyance, laden with the families of Unionists, who find themselves compelled to flee on the approach of the Confederate army, fearing the rebel General will carry into execution his recently made threat to hang every citizen "Yankee" he found in the Kanawha Valley. Hundreds of people who two years ago, were the quiet possessors of large farms, are now driven away from home in a condition bordering on destitution. Unable to remove their farm stock, they are obliged to leave behind them what they depended on for subsistence during the coming winter. Arriving at Gallipolis, or elsewhere, most of them have to seek a charitable home among strangers—a few only, comparatively, have relatives or friends to live with. It is a pitiable sight to see families sent adrift, with their little lots of household furniture, to find a home, they know not where—and all because their father or husband would not renounce his allegiance to the Government of his fathers. The rebels in Western Virginia have declared themselves unsatisfied with anything less than armed resistance to the Federal power on the part of citizens whom they meet in their raids. It will not do to say you have not taken sides either way, or that your sympathies only are with one side or the other. They demand active participation in their cause, and "confiscation," robbery and outrage are the punishments for Federalism. The whites are not the only emigrants from the Kanawha Valley. The negroes have absconded in hundreds, and few less than a thousand have left their disloyal masters to inquire as to their whereabouts and wonder at the answer. The darkies have constructed the most ingenious kind of sailing craft, and in the efforts to elude the rebel advent, which they have learned to dread greatly, have entrusted themselves to the most fragile of home-made vessels. I heard an escaped contraband say, to-day, that he came down the Kanawha fifty miles on a log, but that he would rather drown than remain with his master, who is in Loring's army and is expected home in a few days.

The rebels, the darkies say, have threatened death to the negroes of the Kanawha Valley, whom they accuse of having kept the Federal forces posted as to Confederate movements coming within their knowledge. The acts and orders of some of our Generals ought certainly to acquit the colored race of the charge of acting as spies for us. There is certainly a conflict of opinion on the subject between the Napoleons of the two sides. General Halleck holds that negroes give information to the rebels, and issues his fiat that they be excluded from the Union lines.

THE GUERILLA.

Monday Evening, - - - - - Sep. 29

☞ Owing to the non-arrival of the mail, up to the hour of going to press, we are without the latest Eastern news.

LINCOLN'S DESPERATION, &c.

LINCOLN seems to be getting to the last stages of infamy and despair. Baffled and defeated at every point, he is now writhing under the punishment he promised us. In the last Cincinnati papers is published his fiat, giving notice that he will, on the first of January, 1863, cause to be emancipated all slaves, or persons of African descent, who shall then be in the employ of any person residing in any State still in rebellion against the United States; and that all officers of the army and navy are commanded to show proper respect to the negroes, as freemen, and that they are to assist them in all their endeavors to throw off the shackles of slavery.

To try still more to quench his hellish thirst, he intends violating openly his faith, by sending the prisoners we recently paroled at Harper's Ferry to Minnesota, to fight the Indians, who are giving him much trouble there, and thereby relieving the troops there, so that they can be brought to Washington. Where are the 600 and 300,000 that would sweep out the last vestige of rebellion in thirty days? This looks as if they were sadly wanting. Poor Abe, like a drowing man, has for the last month been grasping at every little straw, but all has been of no avail, and he is now in the last struggles of death, with not the least hope to cheer him in his last moments.

The North seems fully aware of the great loss they have sustained in having to give up the Kanawha, and are free to acknowledge the great importance of its acquisition to our cause. They are bitter against their government for having withdrawn the troops, and acknowledge that we have destroyed in a week what it took millions of money and an army of fifteen or twenty thousand men fifteen months to accomplish. They seem to have no hopes of attempting to retake it this season, at least, as they are now in need of every available man in Kentucky and Maryland; but let them come when and in what force they please, we have no fears but that they would be made to re-enact, in full style, the Lightburn double-quick.

Our brave boys have much to congratulate themselves upon in the good service they have rendered our cause.

☞ THERE will be a meeting of the salt owners here to-day, and it is to be hoped they will fix some established and reasonable price for this much-needed article.

☞ THE STREETS of Charleston are again becoming gay. A great many merchants have re-opened their stores to the public.— Others, however, still keep themselves and their goods shut up in the dark, because they have some scruples about taking Confederate money, &c. We hope they will soon come to their senses, and show that they appreciate their deliverance from the Northern vandals, by immediately opening their stores and offering their goods at the same rate they sold to the Yankees. And it is well here to add, that it is a great wrong and outrage, and speaks poorly for any one to take advantage of his fellow-being in adversity.

☞ PIERPOINT, becoming frightened at the recent audacity of the rebels in polluting his soil, taking his salt works, &c., recently made a visit to Cincinnati in his "Scotch cap," for the purpose of getting aid to punish the audacious and impertinent rebels.— He was met with the happy intelligence, that every available man was needed for the same purpose in Maryland and Kentucky; but as soon as that was finished, he should have immediate assistance to accomplish his purpose. With this happy consolation, he has returned home, no doubt feeling much easier in mind and body.

☞ THE ENEMY, about six or seven hundred strong, came up day before yesterday, to Gen. Jenkins' position, with the evident intention of attacking him, but one shot from our cannon was sufficient to cause the same old "scare" to come upon them, and they immediately skedaddled, and have not been heard from since.

TRIBUTE OF RESPECT.

HEAD'QRS OTEY BATTERY,
Camp near CHARLESTON, Kanawha Co.,
September 23, 1862.

WHEREAS, in the all-wise providence of God, our comrades, Sergeant SPEAR NICHOLAS, CURTIS CHAMBERLAINE, GEO. M. LEFTWICH and HENRY SMITH, have been taken by the hand of death from the ranks of our company, and the scenes in which we loved to mingle with them—

Resolved, That while we bow submissively to the Divine will, in removing from our midst these, our brethren in arms, yet we cannot but mourn the loss of those who were kind in all their relations with us, faithful in the discharge of all their duties, and willing to do and suffer in behalf of their country.

Resolved, That we, as a company and as individuals, do sincerely sympathize with the parents and friends of the deceased, and most cheerfully record our testimony that they fell on the field of battle (Fayette C. H., September 10, 1862,) whilst nobly contending for the rights and liberties of their country.

Resolved, That the foregoing be published in the Richmond *Dispatch* and *Whig,* and copies be sent the families of the deceased.

By order of the Company.

WM. A. HART,
R. G. HARPER,
T. M. NIVEN,   } Committee.
J. R. PERDUE,
A. A. FARLEY.

Sept 29—1t

BY TELEGRAPH.

From the Cincinnati Commercial. Sept. 25.

Gen. Buel's Arrival in Louisville.

LOUISVILLE, Sept. 24, near midnight.— Gen. Buell has just arrived.

Gen. Nelson has just issued an order permitting, to-morrow, a general resumption of business, the issuing of passes to loyal citizens, and the discharge of all enrolled citizens from duty.

The enemy seems to be concentrating at Bloomfield. About 12,000 of them were seen this morning beyond Salt river, on the Bardstown road.

FROM WASHINGTON.

WASHINGTON, Sept. 24.—Intelligence from McClellan is meagre in quantity, and uninteresting. No active movements have occurred within the last day or two, his operations being chiefly confined to reorganizing the army, and other preparations necessary to finishing the rebel rule in Virginia.

Gen. Thomas Morris has been appointed a Brigadier General, and it is reported that he has been assigned to the command of Western Virginia.

THE LONDON TIMES ON THE AMERICAN WAR.

The following is a copy of the leading editorial of the London *Times,* July 22, 1862. In the events that have since transpired, the *Times* will find additional force to its remarks:

Nothing is so melancholy as forced meriment, and sorrow never wrings the heart more bitterly than when affliction or anxiety is compelled to assume the appearance of rejoicing. There is a play of the old dramatist, John Ford, called the *Broken Heart,* in which the heroine is compelled to go through a solemn dance, although in the course of it she is told that her father is dead, that her dearest friend has committed suicide, and that her lover is murdered. She goes through the dance, and drops down dead at the close. Something similar to this must have been the mental torture endured by a large portion of the American public during the festivals and rejoicings which commemorate the 4th of July, the never-sufficiently-to-be-praised and never-sufficiently-to-be-violated Declaration of Independence. How flat, stale and unprofitable must have sounded the conventional eloquence and worn-out enthusiasm which celebrated war and quarrels, the remembrance of which out to sleep in the graves of those who made them! How jarring the music, how pale the fire works, how wearisome the processions to mothers and sisters, to wives and daughters, tormented with the well-grounded apprehension that in the bloody swamps of Virginia were lying those most dear to them on earth—happy, indeed, if dead, but only too probably lingering out the last remnants of existence under the chilling dews of night, consumed by thirst and fever, in all the agonies that wait on the wounded, abandoned by a retreating, and trampled on and disregarded by a pursuing army! Was there no one to ask whether the cause in which all this blood was shed be indeed the cause of independence; whether the North can really identify the grounds of the present quarrel with that

- 103 -

General Loring tells the slaves they have been giving information to the Yankees, and threatens to hang them for so doing. Cuffie has to take the dilemma by both horns.

MACK.

THE RETREAT FROM THE KANAWHA VALLEY.

We are permitted to publish the following interesting letter from an officer of the 34th O. V. I.:

GENERAL U. S. A. HOSPITAL,
GALLIPOLIS, Ohio, Sept. 15th, 1862.

One year ago, to-day, I left Camp Dennison and Cincinnati for Western Virginia.—Little did I then think that the first anniversary of that day would be spent in such a place as this. But here I am, and thankful, too, that I was so fortunate, for many of my comrades fare worse. I have a head not quite so sound as it was five days ago, but if you will agree to make allowances for this, I will give you as correct a description as I can, (together with what I have picked up from those who saw what I did not,) of what was done at Fayetteville, Va., on the 10th of September, and for a few days following.

Fayetteville was occupied last winter by the 23d and 30th Ohio regiments, and under Col. Scammon, had extensive fortifications erected on three different hills, within supporting distance, and the two advanced ones so constructed as to be commanded by the third and largest one. The 34th (1st Ohio Zouaves) and 37th Ohio regiments, with a squad of the 2d Virginia cavalry, occupied the town. Numerous rumors had been in circulation for days, to the effect that the enemy were coming, but we had long since learned to place little confidence in these, and although vigilant guard was kept at all times, no one dreamed of an attack at this time.

At about nine o'clock in the evening, the advance picket on the Raleigh road was driven in, and at the same time word came of an attack on our train at Laurel Creek, five miles out on Gauley Bridge road. Lieut. Colonel Franklin, with four companies of Zouaves, was immediately ordered out to Laurel Creek, by different roads, two companies on each.

Four howitzers were placed in the largest redoubt, and commanded by Lieut. Anderson, of Co. K, 34th. Two large cannon were planted in the earthwork farthest out on the Raleigh road. The six remaining companies of the 34th regiment were drawn up in line of battle, in the rear of the inner redoubt. Two companies of the 37th were sent out on the Raleigh road to skirmish with the enemy; two companies deployed along the south side of the town to guard our left; three companies placed in the middle redoubt, and the remaining three companies held in reserve. Our advanced lines on the Raleigh road were first attacked by cavalry in strong force, a sharp fight ensued, with little loss on our side, but many of the enemy's saddles were emptied.

Our boys fell back slowly, obstinately disputing every rod of ground, until the enemy had come within reach of our outer earth-work. They then planted a battery in and behind a dwelling occupied by a Union family, about six hundred yards from us and opened fire on our ranks. Their guns were vastly superior to ours in calibre and force, and had it not been for the protection

one men had, they would have suffered severely. Cannonading and brisk skirmishing continued until one o'clock. The rebels threw a large force of sharpshooters into a large house about a hundred and fifty yards from our battery, when they knocked off the weatherboarding with their guns and began to work on our cannon. This was soon ended by our throwing a shell into said institution and setting it on fire. That single shell must have killed ten or fifteen rebels.

Soon after this the villains showed themselves in the woods and on the hill on the North side of town. The force here must have been two thousand—one regiment of Georgia sharp-shooters and one battalion of Mississippi Tigers. Our howitzers opened on the woods they occupied with good effect.

The enemy made a great effort to gain the middle redoubt and charged upon it no less than five times, once even gaining the parapet; but, as our boys say, "it was no use," for the gallant 37th now had no idea of giving up their position, and each charge was effectually repulsed, and with great loss to the enemy. Our flag was shot down repeatedly, and completely riddled with bullets.—Once the color-bearer saw his flag fall. He sprang to the top of the embankment, waved it several times, and then planted it in its place, and returned amid a shower of the enemy's bullets.

At about one o'clock our trains were started out on the road towards Gauley Bridge, the ambulances in front. They had not proceeded a quarter of a mile from the town when a heavy fire was opened on them from the woods on the left. Many of the sick in the ambulances were killed and wounded, horses shot, and teamsters badly scared.

This is a demonstration of the enemy on our right to draw the 34th into action. They were marched down the road on the run, and then "by the left flank" up the hill and at 'em. Our boys were soon brought under a very hot fire, but they played the Zouave, firing and loading, lying and then jumping up and advancing a few paces, and hiding themselves again, would let the rebels have another volley. They thus fought for an hour and a half, when it was plainly to be seen that the rebels had too great an advantage over us, being entirely covered by woods. Our forces were therefore withdrawn back to the road, and took up position on the east side, out of line of the enemy, while the little howitzers kept up a continued play on the infested woods.

It was during this last engagement that most of the men killed and wounded on our side were struck. Seven out of eleven officers of the line went down, two killed and one mortally wounded, and not less than eighty out of four hundred enlisted men killed and wounded. Col. Toland had two horses shot from under him. Indeed, it was throughout a very lively time. Cannonading and skirmishing continued on until dark. Lieut. Col. Franklin, after reconnoitering the country out to Laurel Creek, and driving back about 70 bushwhackers, made his way back to camp, where he arrived after dark, his command dusty, thirsty and tired, but all anxious to avenge the death of so many of their comrades, who they now for the first time found had fallen during the day.

All slept on their arms, lying down in ranks. At midnight two companies, I and

K, were ordered to reconnoiter the woods. Company K was deployed, and advanced up the hill. When they had come to within about forty paces of the woods, a heavy fire was opened on them, and many of our boys were wounded and two or three killed. It was here that I got my portion. A Mississippi rifle ball cut a very respectable furrow across my head, which laid me insensible for an hour. From this time I don't know much about what was done, but was told the next morning, when our troops came out, they were compelled to bear a hot fire from the woods during the whole time of passing out. But, as I understand, the tide turned here, and ever since. Although our men retreated, they took occasion at every good place to give their pursuers a warm reception, and without any loss on our side, killed great numbers of them.

The sick and wounded were brought down the river in small boats and in ambulances. It was dusty in the road and hot on the river, but we were glad to get through; and not a man but what hailed the Ohio shore with pleasure. We are pleasantly situated here, and generally in good spirits. Some of the poor fellows will not get well, but the wounds are not generally very serious.

Col. Lightburn has fought the enemy in Seigel's style, and with great success. He has punished their vastly superior numbers at every point, brought off about all our stores, and retreated no faster than convenience dictated.

Col. Lieber, of the 37th, deserves great credit, and a big general's commission, for the manner in which he commanded the two regiments at Fayette, as does Col. Lightburn for his conduct later in the fight down the Valley.

Reports just came in that 640 wagons are safe at Ripley, Va., and Colonel Lightburn faces the enemy. Yours truly, E.

A SPLENDID CHANCE.

A FLYING BATTERY is about to be formed for *Gen. Jenkins' Cavalry Brigade,* to be officered by experienced artillerists, and to be equipped in the most superb style. The Battery is to consist of two three-inch rifled guns, two twelve pound howitzers, (light, such as the Richmond Howitzer Battalion has,) and two mountain rifled guns, to be packed, when necessary, on horses.—Fleet, active horses for the pieces are now being purchased by the Quartermaster of Jenkins' Brigade, and all necessary steps for the procurement of a complete outfit are being taken. Applicants for membership will be required to undergo a medical examination, and must be *young, active and intelligent.* The cannoneers will be mounted, and must furnish their own horses, which will be valued and paid for.

While it is expected to recruit the men from among the mounted companies now forming in this section, transfers can doubtless be procured for a few enterprising men from the regiments and battalions.

The service is a brilliant one, full of exciting incident. *No half-asleep men need apply!* A Recruiting Sergeant may be found for the present at the ORDNANCE OFFICE in Charleston.

which their ancestors maintained against George III. and his Ministers; or whether the parts have not been wholly inverted, or whether the North does not find itself playing the part of the very King whom for eighty years it has held up to the execration of its people as the vilest and most cruel of tyrants?

We know that the North has not succeeded, but can it show any ground to convince us that it deserves success? They cannot submit their cause to be tested by their own principles. Can they point out any others under which their cause will obtain more favor? America has been celebrated, and justly, as the first country that ever based its Constitution on the principles of abstract right and justice. The founders of the Republic maintained the principle of the inalienable rights of man against proscription and authority. Rebels in their eyes were only men reverting to the first principles of natural justice, and sovereigns lost their right to reign as soon as they ceased to contribute to the happiness of their people. These are the stereotyped doctrines of the 4th of July. To be consistent, the Northern States ought now to denounce and punish them as treason.

How are the mighty fallen! A year ago the North went forth to conquer, confident in its numbers, in its vast flotilla, in its crushing artillery, and in its possession of capital, for the moment, at least, entirely without limit. It went forth to fight for empire, and, as men do who seek to conquer and oppress their fellow-men, it trusted mainly in overbearing might, and rested the merits of its cause on the sharpness of its sword. It invaded on every side a territory scantily peopled, supplied with like wealth, without manufactures, without large cities, cut off from the rest of the world by the vast superiority of its antagonist, with nothing to rely on but dauntless courage and resolute endurance. The Southerner was ill-armed, ill-clothed, ill-fed, poorly lodged, and he was encumbered with the most formidable of all hindrances—a slave population of several millions, to whose mercies he had to leave his wife and his child, his homestead and his plantation, when he went forth to fight his battle of independence.— Wherever they could swim the Northern gunboats penetrated, and, so long as they were accompanied by this flying artillery, which also afforded an easy means of transport for all the wants of an army, the Federals proved irresistible.— The time came at last, however, when it was necessary to advance beyond the reach of gunboats, and then, as we in England always predicted, the Federal difficulties began. The Confederates withdrew from before Washington, but the Federals could not follow them, and Gen. Beauregard disappeared from his lines at Corinth, leaving Gen. Halleck quite unable to pursue him.— The great Army of the West has been reduced to inactivity, but the Army of the East has contrived, by marine transport, to place itself on the South-east of Richmond, thus interposing that Capital and the whole army of the Confederates between itself and the remainder of the Federal forces. As if this was not enough, Gen. McClellan disposed his men on a piece of ground divided by three rivers, thus giving every facility for the destruction of his army in detail. The catastrophe has come, as might have been

expected. Almost surrounded by their enemies, the Confederates, moving on shorter lines, had always the opportunity of throwing an overwhelming force on any point which they chose to attack. An advantage once gained was vigorously improved, and after seven days' hard fighting the Federal army is rolled up into a dense mass, the destiny of which every body expects to be very similar to that which has been prematurely announced. After pouring forth blood like water, and fertilizing the fields of Virginia with thousands of corpses, the North finds itself obliged to begin all over again, with credit destroyed, a ruined revenue, a depreciated currency, and an enormous debt.— Nay, as if these were not sufficient, a Republic, based avowedly on the inalienable right of man to personal liberty, to life, and to the pursuit of happiness, and on the principle that Governments are formed for the purpose of establishing these rights, begins to talk of levying 300,000 men by conscription.

Will nothing arrest this frantic and suicidal rage? Is there no one from whom the American people will listen to the words of truth and soberness? We know that counsels of moderation, ever distasteful in themselves, are doubly distasteful when coming from us; but we can scarcely believe that the infatuated multitude will remain as blind to the teaching of facts as they have hitherto been deaf to the voice of well-meant expostulation. What proof do they yet require that they are embarked on a fatal and ruinous cause? Their wealth is turned into poverty, their peace into discord, their prosperity into wretchedness; the power in which they gloried is effaced; society is torn in pieces by the hands of its own members; law is trampled under foot, and the country is fast falling into anarchy, the only refuge from which is despotism. We do not scruple to say, that we shall rejoice if the worst anticipation be realized—not from any ill will to the North, but because we see in the failure of its efforts to subjugate the Southern States the only prospect—we had almost said the only possibility—of peace.

There is something pathetic, when compared with the language to which we have been accustomed, in the extreme thankfulness with which New York hails the escape of the army from absolute perdition. We hear no more of "strategic movements," of bets that McClellan would be in Richmond in a week. Truth and nature have at last found utterance, and the language of empty swagger and wilful falsehood is thrown aside. "Could the army once make a stand, be permitted a brief interval of rest, and time to throw up entrenchments, under the protection of the gunboats, on the James River, all would be well. On Wednesday and Thursday there was no fighting. We cannot exaggerate the importance of this fact, nor exult too much over it; it is the salvation of the army; therefore, though we have, perhaps, lost men by thousands and guns by hundreds, we announce the news with gladness." We earnestly hope that this soberness of tone and humbleness of expectation is an indication of the first step in a change of public opinion, which may induce the North to shake off the sanguinary dream of conquest and empire, and return to a due estimation of its own interests, and the rights of those whom it has been its futile ambition to trample under foot.

GENERAL ORDER.

HEADQUARTERS, DEPT. OF WESTERN VA.,
CHARLESTON, VA., Sept. 17, 1862.

General Order No. —.

The com'dg General feels deeply sensible of his obligation to treat all persons in arms against the Confederacy, and who may fall into his hands, with the utmost humanity required by Christian charity and the usages of war. In like manner all citizens who obey the laws and repudiate their treason, will be treated with the clemency declared in the Commanding General's Proclamation. He has heard, with deep mortification, that in a single instance this rule has been departed from by the unauthorized order of one of his officers; but the wrong done will be promptly punished and redressed.

All persons who have received arms of the public enemy are invited to bring their arms into camp, and if they choose, take service in the defence of the country. No punishment will be imposed on such persons.

By order of
MAJ. GEN. LORING.
H. Fitzhugh, Chief of Staff.
Sep 26-tf

GENERAL ORDER.

HEADQUARTERS, DEPT. OF WESTERN VA.,
CHARLESTON, VA., Sept. 14, 1862.

General Order No. —.

The Commanding General congratulates the Army on the brilliant march from the Southwest to this place in one week, and on its successive victories over the enemy at Fayette C. H., Cotton Hill and Charleston. It will be memorable in history, that, overcoming the mountains and the enemy in one week, you have established the laws and carried the flag of the country to the outer borders of the Confederacy. Instances of gallantry and patriotic devotion are too numerous to be specially designated at this time; but to Brigade Commanders, and their officers and men, the Commanding General makes grateful acknowledgment for services to which our brilliant success is due. The country will remember and reward you.

By command of
MAJOR GENERAL LORING.
H. Fitzhugh, Chief of Staff.
Sept 26-tf

GENERAL ORDER.

HEADQUARTERS, DEPT. OF WESTERN VA.,
CHARLESTON, VA., Sept. 15, 1862.

General Order No. —.

All public Stores, Horses, Wagons, and property of every description, captured by the Army, or in possession of private citizens, will be handed over to the Quartermaster. All plundering of such property will be severely punished. The Commanding General learns that great waste has occurred by want of attention to the law in this respect, and by appropriation of such property in the Army.

By order of
MAJ. GEN. LORING.
H. Fitzhugh, Chief of Staff.
Sept 26-tf

GENERAL ORDER.

HEADQUARTERS, DEPT. OF WESTERN VA.,
CHARLESTON, VA., Sept. 18, 1862.

General Orders No. —.

On and after this date, all persons arriving at this place will report at once to Major Thomas Smith, Provost Marshal.

By order of
MAJ. GEN. LORING.
W. B. Myers, A. A. G.
Sep 29-tf

General GILLMORE has assumed command of the forces in the Kanawha Valley. Order is already coming out of the chaos into which matters were thrown by the retreat from Gauley.— It is quite apparent that a military head is now at work. We see fewer drunken, disorderly soldiers, roving about town, officers riding furiously in pursuit of nothing, no longer render our public thoroughfares dangerous to footmen.— Interminable trains of wagons, pass without any longer blocking up the roads, and in short every thing begins to wear the appearance of war, not in Gallipolis but on Kanawha. Forces are coming in daily, and are quietly taking up their positions. But a short time must elapse, when the Secesh will have to take the other end of the road on the double quick. It cannot be done a day oo soon. Our Union friends from the Valley are anxious to return home if they should be so lucky as to find one left them. They deserve to be protected and we feel satisfied they will be. They will not be annoyed either by finding the property of their Secesh neighbors guarded by Union soldiers. That game is played out.— Secesh property will be decidedly below par hereafter. The Government owes it to the true and loyal men of Kanawha that their valley shall be rid of these pestilent fellows. Let the salt furnaces either be stopped effectually or be placed in hands of loyal men. These salt boilers have done more to protract this rebellion, by furnishing niggerdom with a supply of salt, than any other class of men in the State.— It is idle to say they could not avoid it, that the rebels took the salt, &c. Did they take any of the owners? Have

The Gallipolis Journal, Oct. 9, 1862

Army Correspondence.

LETTER FROM THE KANAWHA.

For the Gallipolis Journal.

CAMP NEAR PT. PLEASANT, VA.,
Oct. 19th, 1862.

ED. JOURNAL:—

At last the tedium of inaction has been broken. Something new, something long desired, seems to have been inaugurated. Can it be a forward movement? We shall see. All is life and animation now. Tents are being struck, and already the work of destroying such property as cannot be removed has commenced. But we are not to move till morning, so the ground, with nothing to shelter us from the chilly frost, must be our resting place to-night. But we are going to move, and all is good cheer. Many are the conjectures as to where we are about to go; some suppose to Kentucky, others up the Ohio, and others up the Kanawha, all having good reasons for their beliefs. We settle down upon the conclusion that it matters not, we are in for it now, and we must learn to wait as well as conjecture. Little is the sleep that many of us enjoy on account of the many more who are too jubilant to rest.

UP-HILL WORK.

Highly Important from Western Virginia.

Desperate Fighting—Two Regiments Attacked at Fayette by 5,000 Rebels.

GALLIPOLIS, Sept. 14! On Wednesday, the 10th inst., a column of the enemy, about 5,000 strong, said to be under command of General Loring, the first notice of whom was in our rear, between Fayette and Gauley, made an attack on our forces encamped at Fayette, consisting of the 34th and 37th Ohio Regiments, numbering about 1,200 men, under command of Colonel Siber, when a desperate fight ensued, lasting till dark.— Our forces cut their way through, reaching Gauley Bridge during the night, having lost about 100 killed and wounded, mostly of the 34th Ohio.

In the mean time another column of the enemy approached Gauley Bridge, on the Lewisburg road, under the command of Cerre Gordo Williams, cutting off the 47th Ohio, two companies of the 9th Virginia, and one company of 2d Virginia Cavalry, who were at Somerville. Nothing since has been heard of them. Under these circumstances, Colonel Lightburn's front flank and rear being threatened by an overwhelming force, was compelled to evacuate Gauley, which he successfully accomplished on the morning of the 11th, after destroying all the government property that he was unable to bring away. He accordingly moved down the Kanawha, in two columns, one on each side of the river, reaching Camp Piatt on the afternoon of the 11th, skirmishing the whole way. Here he massed his troops on the north bank of the Kanawha, but being hard pressed by the enemy he retreated during the night, reaching Elk river, just below Charleston on Saturday morning. He made another stand on the lower bank of Elk river, and a desperate battle ensued, lasting from ten A. M., till dark. Our forces shelled and destroyed Charleston, two houses only being left.

The result of the fight is unknown, nothing having been heard from Col. Lightburn since 6 o'clock on Saturday evening. Up to that time our troops held their own, and were punishing the enemy severely.

We understand that our forces completely destroyed all the salt works. Col. Lightburn brought an immense train of 600 loaded wagons safely to Elk River. The retreat to Elk River was conducted in good order.

Great anxiety is felt for the safety of our forces, as well as of Point Pleasant and Gallipolis. The militia are flocking here from this and surrounding counties. This border is in great danger. The enemy's force is represented as being 10,000 strong, with a proportionate force of artillery.

The Gallipolis Journal, Sept. 14, 1862

Uneasy to learn the actual condition of affairs with Lightburn's command, I determined to reach Gallipolis the same night. Our horses had been left behind, and being then dismounted, we took passage in a four-horse hack, a square wagon on springs, enclosed with rubber-cloth curtains. Night fell soon after we began our journey, and as we were pushing on in the dark, the driver blundered and upset us off the end of a little sluiceway bridge into a mud-hole. He managed to jump from his seat and hold his team, but there was no help for us who were buttoned in. The mud was soft and deep, and as the wagon settled on its side, we were tumbled in a promiscuous heap into the ooze and slime, which completely covered us. We were not long in climbing out, and seeing lights in a farmhouse, made our way to it. As we came into the light of the lamps and of a brisk fire burning on the open hearth, we were certainly as sorry a military spectacle as could be imagined. We were most kindly received, the men taking lanterns and going to our driver's help, whilst we stood before the fire, and scraped the thick mud from our uniforms with chips from the farmer's woodyard, making rather boisterous sport of our mishap. Before the wagon had been righted and partly cleaned, we had scraped and sponged each other off and were ready to go. . . .

Excerpt from *Military Reminiscences of the Civil War*, by Jacob D. Cox on his return to the Kanawha Valley in October 1862.

TO THE LOYAL CITIZENS
OF THE
KANAWHA VALLEY.

The 8th REGT. VA. VOL. INFANTRY has returned to the Kanawha Valley. This regiment served as the advance guard of Fremont's army in the Valley of Virginia, and was several times complimented by Gen. Fremont in General Orders, for its gallantry in action and behavior on marches. Afterwards it became a part of Sigel's corps, and after performing an honored part in the bloody campaigns of the Rappahannock, and participating in the Bull Run fight, it hastened back to Western Virginia, eager to assist in driving the rebel hordes from their homes and yours.

Having been in service nearly eighteen months, its ranks are much thinned. Half its term of service has expired, and you are now appealed to, to come forward and volunteer to fill up the ranks for the rest of the term.

All who wish to enter the service of their country, with veteran troops, and under officers of experience and tried courage, have now a glorious opportunity. You will receive the same pay, bounty, clothing, &c., as in any other regiment. Come on, then, loyal men of Virginia! Range yourselves side by side with your friends and brothers, and drive the ruthless rebels from our soil.

.... BOOTH, Dec. . . 1862.

A poster issued at Coalsmouth (St. Albans) in December 1862 recruiting new members for the 8th Regiment of Virginia Volunteer Infantry (Union). WVC

U. S. COMMISSIONER'S OFFICE, Charleston, Va. *August 15th* 1862.

This is to certify that *Francis Davis* of *Kanawha* county, Virginia, has

taken the oath of allegiance, and is entitled to the protection of the Government of the United States, but will be de-

nounced and punished as a TRAITOR for any violation of said oath.

E. W. Newton U.S.   Commissioner

No. *Office*   Head-Quarters, *Provost Marshals*

Charleston Va. DEC 29th 1862

All guards, lines, post, stations will pass safely *William*

E. Chilton

from *Charleston* to *Mouth Big Sandy*

DESCRIPTION.

Complexion.	Hair.	Height Feet—Inches.	Eyes.	Age.	Whiskers.
Light	*Dark*	*5 - 10*	*Blue*	*35*	*Light &c*

This pass is given with the understanding that if the party receiving it be found hereafter in arms
against the government of the United States, or aiding or abetting its enemies, the penalty will be death.

Ernest Schache
Major and Provost Marshal

Edge of the Cauldron
Putnam County

The Civil War in Putnam County
by IVAN M. HUNTER

When the Civil War broke out, Putnam County was actually a border county, in a border state. This meant that geographically the county was near the dividing line between the North and South. Neighbors, friends and even families were divided over the deep issues. The area became a scene of wild excitement and great confusion. Issues were argued, tempers flared and young men left home to enlist for the cause they held to.

The first clash of any importance between the gathering forces in the area came on July 14, 1861. Four companies of the 2nd Kentucky, under Lt. Col. George Neff, met a Confederate force of considerable size at Barboursville. The fight took place at the bridge across the Mud River. When the citizen militia got their first smell of powder and heard the whistling Minie balls, they hastily beat a retreat and then left the fighting up to Capt. James Corns and his mounted Sandy Rangers, better known as the "Blood Tubs" because of the bright red shirts each soldier wore. Facing superior numbers and better equipped feds, the rangers were forced to withdraw after a short but bitter fight. The Rebels lost one man killed, Absolom Ballinger of Milton, while Union troops lost five killed and 18 wounded.

Two days later the spotlight of battle shifted to Raymond City and its twin village of Poca. It was then that a detachment of General Wise's army engaged an advance party of Union soldiers near the mouth of Poca River. Old-time residents could recall that Wise had mounted a cannon on the top of the hill over-looking Raymond City, to fire a warning shot when the Federals were sighted advancing up the Kanawha. The booming of the gun sounded early on the morning of the 16th as the pickets were driven in. Col. John Clarkson, with Brooks and Beckett's troops of horses, numbering about 120, waited behind hastily erected breastworks. Two six-pound cannons were trained in the direction of the advancing Yankees. The report of the engagement listed eight Union soldiers killed, with only the loss of one horse on the Confederate side.

The next day, July, 17, 1861, the main forces of the opposing armies met at the mouth of Scary Creek, across the river from present-day Nitro. At about 1:30 p.m., the Confederate pickets of Captain Barbee's Border Riflemen (Putnam County troops) were driven in from their position on Little Scary Creek. Soon the Union forces moved into position facing the Confederate line, above and bordering on Scary Creek.

When the Yankee Cavalry broke into sight at the Robert M. Simms farm on the opposite ridge, Capt. George S. Patton, commanding the Rebel artillery, ordered both pieces to open fire. (Patton was the grandfather of "Blood and Guts" Patton of World War II fame.) This caused the Union Cavalry force to withdraw in haste to await the arrival of the main body.

The Battle of Scary Creek, for the most part, was fought at long range with rifle and artillery fire from each side of the creek. With the arrival of two infantry units the battle began in earnest. An hour elapsed with little change in the situation. Bullets whined through the trees, the boom of the field pieces echoed from the hillsides. Now and then a horse would scream as a piece of lead thudded into its flanks.

About three o'clock the Ohioans attempted to charge the bridge. But the Rebels held their fire until they were almost over and then opened up with everything they had. This was too much for the Federals and they broke ranks and retreated in wild disorder.

The Union forces however regrouped quickly and made another charge at the bridge. This time they made it across but were unable to hold their hard won position and were forced to withdraw, leaving Lt. Col. Jesse Norton of the 21st Ohio, on the field shot through both hips. At the same time Capt. Patton was shot in the shoulder while trying to rally his troops, while exposed to the enemy.

Capt. Albert G. Jenkins assumed the command when Patton was wounded. The time was late afternoon and the Federal troops were forming another run at the bridge. Both of the Confederate guns had been knocked out and Lieutenant Welch, an Artillery officer, had been killed. Some of the volunteers had taken off, and it appeared as if the Federal troops were gaining a tremendous advantage. Just as another Union charge was about to begin, Capt. James Corns who had been stationed atop Coal Mountain galloped into the fight with his Sandy Rangers. Also right on his heels came Captain Thompson with his Kanawha militia and one piece of artillery.

The "Blood Tubs" swung into action singing a ballad, "Bullets and Steel." One survivor remembered years later that they could be heard above the roar of the battle. This action turned the tide of battle in favor of the men of the south. The Federal troops, thinking that Wise's reinforcements had arrived from the Kanawha Two-Mile Camp, broke ranks and retreated toward the Morgan farm.

For some unexplained reason, Captain Jenkins, seeing the Federals falling back and probably thinking they were regrouping for another assault, ordered his forces to withdraw. For a short time the battle field was deserted by both sides. Col. Frank Anderson from General Wise's staff had been commanding two rifle companies on the left flank and he quickly sized up the situation. Taking command, he quickly returned to the field and claimed the victory. The overall losses of the Federal troops were 15 killed, 11 wounded and 7 captured. The Confederate casualty list included 4 killed, 10 wounded, and 2 captured.

Directly across the Kanawha River from the mouth of Scary Creek, one family listened to the firing with great concern. Besides having a river-side view of the battle and being in the line of fire from stray bullets, the Mason family had a closer interest in the engagement. Two brothers of the family, Thomas L. Mason and William M. Mason, got through the battle at Scary unscratched only to lay down their lives later in the same war. Thomas was killed at the Battle of Lewisburg shortly after, and William fought until 1864 only to be struck down at Leestown, Virginia. Ella Mason often recalled the terrifying hours of the conflict as she and the other children huddled behind stout furniture during the height of the battle across the river.

The Battle of Hurricane Bridge was fought March 28, 1863. It was known to the Confederates that at Pt. Pleasant there was a large amount of Federal army supplies, and a large number of horses. Gen. Albert G. Jenkins, commanding a cavalry brigade at Dublin Depot, on the line of the Virginia and Tennessee railroad, determined upon a raid across the mountains and down the Kanawha Valley for the purpose of capture. On March 20th a detachment of eight hundred men, partly made up of the 8th and 16th Virginia cavalry regiments, commanded by General Jenkins in person, with Dr. Charles Simms of Putnam County, as surgeon, began the two-hundred-mile march over the mountains, despite bad roads and weather. March 27, the column reached Hurricane Bridge, where a Federal force was stationed, consisting of Co. A, under Captain Johnson; Co. B, Capt. Milton Stewart; Co. D, Capt. Simon Williams; all of the 13th W. Va. Infantry, and Co. G, of the 11th W. Va.

Early in the morning of the 28th, Maj. James Nowling, of the Confederate forces, under a flag of truce, reached the headquarters of Captain Stewart, the senior Federal officer, and demanded an unconditional surrender. Stewart refused to comply, and Major Nowling left, remarking that "within thirty minutes an attack will be made," and he made good his threat, and the sound of musketry was heard within that time. It was returned with much effectiveness, and for five hours the engagement continued. The Confederates then withdrew and resumed their march toward the mouth of the Kanawha. The Federals suffered several casualties, including Ultmas Young and Jesse Hart, both killed.

The Battle of Winfield was a night engagement on Oct. 26, 1864. Capt. John M. Reynolds, commanding Co. D, 7th W. Va. Cavalry, was sent to oc-

1850 MAP
of
COALSMOUTH and THE MILITARY BOTTOMS

Original Map made by C. B. Shaw, C. E., for the Commonwealth of Virginia in connection with a survey made for The Covington and Ohio Railroad.

Published with permission of the C&O Historical Society, Clifton Forge, Virginia.

ENHANCED by WILLIAM D. WINTZ

cupy Winfield to give projection to river transportation. There he constructed rifle pits, traces of which were still visible in 1892. Late in October Col. John Witcher, of the Confederate army had regiments along the Mud River, and hearing that the Federals had fortified Winfield, he decided to attack them. The attack was carried out at night with 400 men divided into two groups. The force commanded by Captain Thurmond reached the center of the defenses first at Ferry and Front streets. They were met by strong fire from the Union troops and Colonel Thurmond fell mortally wounded. The Confederates were forced to fall back and after capturing several horses, they withdrew back to Mud River, leaving the Federals in possession of the town.

The next day, under a flag of truce, Capt. W. R. Bahlman came into town carrying a coffin on his back and he personally buried Capt. Philip Thurmond in an unmarked grave, behind the old Hoge house.

The Last Bivouac

When Captain Patton came to Coalsmouth to recruit volunteers for the Confederate army, two young men from across the river were among the first to sign up. They were 16-year-old James Rust and his friend Henry Gregory who was 21. They both lived on Ferry Lane opposite the mouth of Coal River where Gregory ran a grocery store for his widowed mother. James Rust was a student at Rev. Nash's school for boys at Coalsmouth and he also worked on the ferry operated by his father Samuel Rust. Not only were they good friends, Henry Gregory had married James Rust's sister Sarah. They enlisted on the same day and both were assigned to Company A of the 22nd Virginia Infantry.

Before he left for the army, James Rust and some members of his family hiked to the top of the mountain on the north side of the Kanawha. It overlooks the mouth of Coal River and Rust was so impressed with the view that he remarked that if he should be killed in the war that was where he wanted to be buried.

A year later, while the Confederates were making a victorious drive to retake the Kanawha Valley, James Rust was killed during the Battle of Fayetteville. After the battle the Federal forces began a massive retreat that continued all the way to the Ohio River.

Since Henry Gregory was still in the same company he was given permission to transport James Rust's body home with the advancing troops. Three days later forward elements of the 22nd and 36th Va. Infantry arrived at Coalsmouth.

Victoria Hansford wrote: "While our hearts were overflowing with joy for the return of our friends, we sadly missed those who would come no more. There were four in all, Charlie Turner, Theodore Turner, Thornton Thompson and Jimmie Rust. I climbed to the top of Rust Hill to see him buried, it was a sad,

sad time. I remember my father sympathized with them so deeply. I can see now he felt for them more than we young folks did. Life was bright for us that even death could not cast a shadow for long."

As fate would have it, less than a year later Henry Gregory was also killed at the Battle of White Sulphur Springs. Eventually his body was also returned home and buried beside his boyhood friend and comrade-in-arms.

As late as 1886 a post office was located at Ferry Lane named Gregory. Ferry Lane was the same as present-day Walker Street in Nitro. Henry Gregory's younger brother William became a well-known steamboat captain. He also married Sarah, his brother's widow, and they continued to live at the old home place.

In Nitro on a high hill overlooking the valley where the Coal River flows into the Great Kanawha, a lonely little graveyard stands weed infested and seemingly forgotten by the 20th century citizens of the valley. Jimmy Rust and Henry Gregory are still together as they had been boyhood friends and later comrades in arms. Here is written the true story of war. Two young men who died before their lives had hardly begun.

This house was built by Lewis Bowling around 1848. At the start of the war Bowling had 125 slaves, whom he attempted to march back to old Virginia. Most of them, however, managed to escape on the way. Both armies occupied the house, using it a different times as hospital and headquarters. The house, near Nitro, survived until the early 2000s.

The Pass

While camped at Buffalo I was given a pass, which afterwards passed me through any lines in the Confederate States. There were three brothers of us, one of whom must go home (to keep the farm going). As I had this pass I really did not belong to the army as the other two did, therefore, it fell on me to go. It was the hardest thing I ever had to do. I told the boys I would come back and that I would be with them when they dug the last ditch. I kept the faith and I was there. This I consider the grandest, proudest feat of my life.

Our company was organized before the war began, and numbered one-hundred and sixty men. I was defeated by Captain by only one vote. When I requested this pass the Board of Examiners, consisting of General John McCausland (then Lieutenant Colonel), David S. Ruffner, Sam Miller, and Dr. Watkins, objected giving it to me. Our Captain, A. R. Barbee, however, told them if I went home anyway dissatisfied and my comrades knew it, half of the company would go with me.

I was a very tall man, second sergeant on the extreme right. I was the first man called in after the officers to take the oath for the State of Virginia and to be examined physically. It was at this time I was compelled to request the pass, which deprived me of the glory of being a soldier. I have often wished I had never received it. I might have died a miserable death in a hospital, or on the battle field, but if I had my life to live again, I would risk it.

Our company was composed of men from two counties, Putnam and Kanawha, about eighty from each. Putnam got the two highest officers, Captain and First Lieutenant. My brother Jim was unanimously elected to fill the latter office. Kanawha claimed the right of filling the three next highest offices, Second Lieutenant, Quarter Master, and Orderly Sergeant. Then coming back to Putnam I was elected next highest, which was Second Sergeant. In voting the two candidates stood side by side and the men lined up behind the man of their choice. The first man who came up behind me was an Irishman, who said: "Faith and be jabbers, I will vote for Jahn, because he has got a great stand." Looking back at the line, I saw all the neighbor boys but one. Why he voted against me I never knew but voting as he did prevented a tie, in which me life, no doubt, would have been a very different one.

When I say the pass to which I have referred deprived me of being a soldier, I mean it put me under a disadvantage from the first. In other words, it prevented opportunities which otherwise would have been mine. My grandfather, Colonel J. M. H. Beale, wrote us that our only sister would live but a short time as she was dying of consumption. She was crying to see her father and at least one of her brothers before she must die. He also feared our mother would lose her mind, as things were in very bad shape at home and we would soon have nothing to come back to. It was then that we three brothers got together and decided that I should go, since our father positively refused to go unless one of us accompanied him. In his letter, my grandfather informed us that we would be permitted to return home by the Union Commanding Officer, General Cox. He also said we would not be required to take the Oath of Allegiance to the United States Government.

This little story was written by Confederate soldier John Morgan. It appeared in his book *The Last Dollar*. He was the father of Sid Morgan, the "Sage of Putnam County."

The First and Second Battles of Hurricane Bridge

It is a little known fact that there were actually two battles at Hurricane Bridge. In both encounters the same man was one of the leaders of the forces gathered there. He was Albert Gallatin Jenkins, the brilliant lawyer, statesman, and battlefield commander.

The first battle of Hurricane Bridge in 1856, like the second, was an all-day skirmish and many of the same basic issues were involved. The main difference was that the one in 1856 was a heated political conflict while the second, in 1863, was a brisk military skirmish during the Civil War.

The following article taken from a biography of Jenkins by then-Congressman Ken Hechler describes the battle of the politicians at Hurricane Bridge. The story appeared in the *Herald-Advertiser*, Huntington, W. Va., August 13, 1961.

> The high point of the campaign came at Hurricane Bridge in Putnam County on Sept. 20, 1856, when Jenkins engaged Congressman Carlile in a verbal toe-to-toe slugfest. A tremendous crowd assembled for a noon-day barbecue on Saturday in a beautiful grove near the town. After the barbecue, the assembled gathering retired to a meeting house to listen to the speakers. Congressman Carlile led off with a speech of an hour and a half, charging the Democratic Party with responsibility for the "slavery agitation" in the country. Jenkins, tall, dark and articulate, then talked for an hour and a half himself, polishing off his address with a series of questions to Carlile which he dared the Congressman to try and answer. In a rebuttal of three-quarters of an hour, Carlile tried to blunt the force of Jenkins' blows, and went on the defensive by trying to answer the questions. This gave an opportunity for Jenkins to spend his final three-quarters of an hour in a series of smashing conclusions to leave the best final impression.

Editorially, the *Kanawha Valley Star*, issue of Sept. 30, 1856, summarized the Hurricane Bridge encounter in this way:

> Mr. Carlile has heretofore been regarded by his friends in this quarter as the Big Gun of Know Nothingism: that Big Gun is now completely spiked.
>
> "The enthusiasm of the Democracy was perfectly unbounded, and toward the close of Mr. Jenkins' speech surpassed everything that we have ever witnessed.
>
> "To judge of the ability and talent of Mr. Jenkins, it is only necessary for you to hear him. He is a young man of superior intellect, and is likely to make a statesman of the first order."

A letter to the editor of the same newspaper claimed that "it was admitted on all hands that the young champion of Democracy had vanquished the Goliath of Know-Nothingism."

When Jenkins was at Hurricane Bridge the first time in 1856 he could not have imagined that he was destined to return seven years later as a Confederate General. By 1863 fires that had been kindled at meetings such as the one at Hurricane, had been blown into the raging holocaust of the Civil War.

Hurricane Bridge battle site, at the junction of US 60 and state Route 34, just south of Hurricane. This was the scene of a battle on March 28, 1863. The Confederates knew that at Point Pleasant there were quantities of Federal army supplies and a large number of horses. Gen. Albert G. Jenkins, commanding a cavalry brigade at Dublin Depot, on the Virginia and Tennessee railroad, proposed a raid across the mountains and down the Kanawha Valley to capture the supplies. On March 20 a detachment of 800 men, partly made up of the 8th and 16th Virginia cavalry regiments, commanded by General Jenkins himself, began the 200-mile march over the mountains, despite bad roads and weather.

On March 27 the column reached Hurricane Bridge, where there was a Federal force consisting of Co. A, under Captain Johnson; Co. B, under Capt. Milton Stewart; Co. D, under Capt. Simon Williams; all of the 13th W. Va. Infantry, and Co. G, of the 11th W. Va.

Early in the morning of the 28th, Maj. James Nounnan, of the Confederate forces, under a flag of truce, reached the headquarters of Captain Stewart, the senior Federal officer, and demanded an unconditional surrender. Stewart refused to comply, and Major Nounnan left, remarking that "within 30 minutes an attack will be made," and he made good his threat. Soon the sound of musketry was heard; fire was returned with much effectiveness, and for five hours the engagement continued. The Confederates then withdrew and resumed their march toward the mouth of the Kanawha. The Federals suffered several casualties, but none were reported by the Confederates.

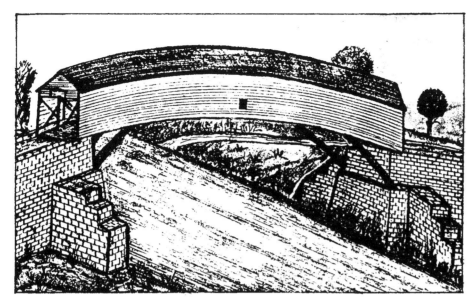

A soldier's concept, drawn from memory, of the first bridge built across the Coal River in 1832 to accommodate the new James River & Kanawha Turnpike. In July 1861 this bridge was destroyed by retreating Confederate forces and not rebuilt for 10 years. The existing Main Street bridge is at this same location. PHC

Ravenswood, located on MacQueen Boulevard in St. Albans, traces its history to the 1830s when a frame dwelling was constructed on the property. The present brick structure was built in 1833 by Francis Thompson. The house had several owners until Judge J. B. C. Drew purchased it in 1897. Trading on the name Ravenswood, he fabricated a story that Edgar Allen Poe wrote his famous poem "The Raven" in the home. Extensive remodeling was undertaken in 1914.

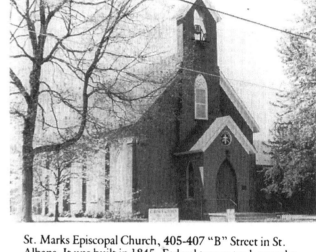

St. Marks Episcopal Church, 405-407 "B" Street in St. Albans. It was built in 1845. Federal troops took over the church in 1863, tore up the floorboards for firewood and used the interior as a stable. In 1915 the Federal Government made a small restitution to the congregation for the damage.

Morgan's Plantation Kitchen on US 60 at the St. Albans Riverside Park. This small log building, which was originally located on John Morgan's farm in Putnam County, was used as a hospital after the Battle of Scary Creek.

A. S. "Sid" Morgan Stories

The Steamboat Ambush

There were many stirring events that took place along the lower Kanawha during the Civil War. Troops on both sides surged back and forth and the people were about evenly divided in sentiment. In some instances, it was father against son and neighbor against neighbor. This led to some exciting and unusual occurrences.

Early in the war near Fraziers Bottom, the Union steamboat *Victor No. 2*, with Stape Wright at the wheel, was going down the river when it was fired on by a band of Rebels. It was the day after the Battle of Hurricane Bridge and Colonel Jenkins' forces were on their way to Pt. Pleasant. They had seen the boat coming and had hidden along the bank in hopes of capturing it with its load of passengers and baggage.

When it came into range, the Confederates fired across its bow and hailed an order to pull over. Instead of pulling up, Captain Wright rang for full speed ahead and hunkered down behind the wheel and ran through a gauntlet of rifle fire. The glass was shattered all about him and the pilot house was riddled with Minié balls but he brought her through with no one hurt.

Some time later, the Union authorities at Gallipolis presented Captain Wright with a fine new rifle with his name engraved on it in gold. The award was made in appreciation of his courage and is still in the possession of his descendants who live in Winfield.

The Battle of the Thrashing Machine

It is a well-known fact that war is no laughing matter, but an incident happened in Poca Bottoms during the Civil War that would have to be classed as a comedy of errors.

It was in the early fall and a reaping machine was in operation on Dr. J. J. Thompson's farm across the Kanawha from the mouth of the Pocatallico River. The big reaper (not to be confused with the thrashing machine, referred to in the title, which was not invented until around 1888) was being used to harvest the large wheat crop that grew in the fertile river bottoms. It was not unusual for thirty or forty men to be working around one of these machines when it was in operation.

Well, on this particular day, they were in the process of moving the machine to a "new set," or to another field. The horses were out in front pulling and the men were all walking along behind their bright pitchforks over their shoulders, gleaming in the sun.

Just about that time, a steamboat passed down the river with its pilot house barely visible through the tree tops that grew along the bank. The pilot, catching glimpses of the horses and the men with their shiny pitchforks over their shoulders, thought it was a detachment of Rebel soldiers in the area.

On reaching Winfield, he quickly reported what he had seen to the Captain of a Union detachment encamped there. As soon as they could be dispatched, a troop of cavalry was on the double, headed back up the valley. Galloping most of the seven miles over the dusty road on such a hot day, their horses were about gone when they reached the wheat fields.

Arriving on the peaceful harvest scene, they charged all around the outlying brush trying to find something to shoot at. They soon had to give up and dismount to give their horses a rest before they collapsed.

Just about the time the soldiers had decided there must have been a mistake and that there had been no enemy, a shot rang out on a nearby hillside. Clamoring back into their saddles they went charging off again in the direction from which the shot had come.

After milling around in the woods for about twenty minutes, a sergeant emerged with John Morgan and a Horton boy. On turning them over the captain reported that at least he had captured a man with a gun. It was soon determined that the boys were only squirrel hunting and the captain, being very courteous, apologized and let them go.

Turning to his men, with the traces of a weak smile on his face, he gave the order to mount up. The hot, tired soldiers and horses then began a slow retreat back to Winfield.

Dr. Thompson, in later years when telling this story, always referred to it as the "Battle of the Thrashing Machine."

An intersection on the highway of American History the Great Kanawha River flows at Point Pleasant, W.V., into the Ohio. In 1774 the Battle of Point Pleasant was fought here between the Indians under Chief Cornstalk and the Virginia militia under Andrew Lewis. It has been called by some the first battle of the American revolution. During the Civil War the beautiful Ohio was a thoroughfare for Federal forces. Ironically the grave of C.S.A. Gen. John McCausland lies on the hill to the right, situated some say so the old man can keep watch on the "bluecoats."

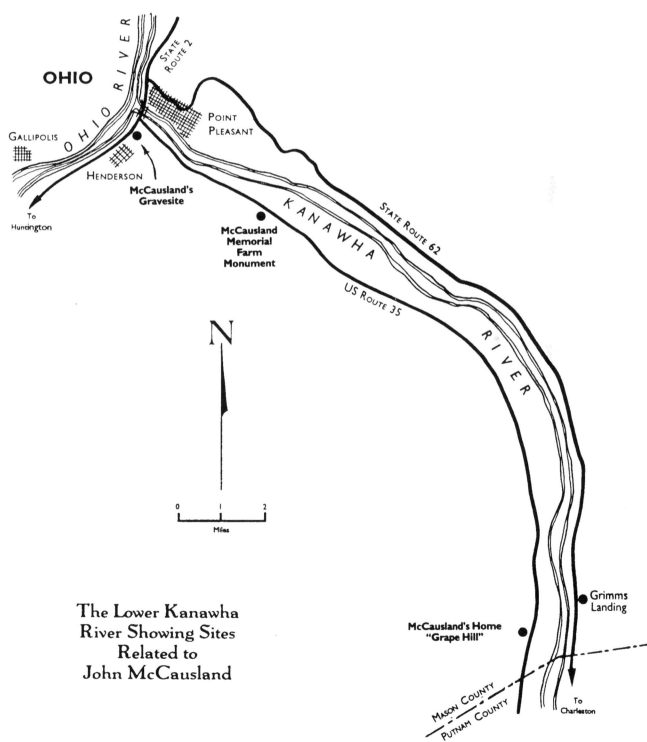

OHIO

OHIO RIVER

STATE ROUTE 2

POINT PLEASANT

GALLIPOLIS

HENDERSON

McCausland's Gravesite

To Huntington

McCausland Memorial Farm Monument

KANAWHA

STATE ROUTE 62

US ROUTE 35

RIVER

N

0   1   2
Miles

McCausland's Home "Grape Hill"

Grimms Landing

MASON COUNTY
PUTNAM COUNTY

To Charleston

The Lower Kanawha
River Showing Sites
Related to
John McCausland

Yankee Caper in Putnam County

by FORREST HULL

The war was fun in '61. The wholesale killing had not started and army life had its pleasant aspects.

The boys from the farms and factories of eastern Ohio who had listened to the speeches and had enlisted at the little county seats found the war, so far, to be something like a hunting and camping-out trip with all expenses paid. It was all jolly good fun.

From Gallipolis and Point Pleasant it was but a short distance to the seat of war, the Kanawha Valley, which was reached after a pleasant steamboat ride. The Federal Army had sent the rebels skedaddling eastward. Only a few bushwhackers remained. And a civilian population composed of a few salt tycoons and poor dirt farmers on lonely creek farms. As the summer of 1861 came to a close the real war seemed remote and far away except for some minor incidents and alarms.

To keep the folks back home in Ohio advised of the life of a soldier, a correspondent of *The Cincinnati Gazette*, a member of Piatt's 34th regiment, known as "Piatt's Zouaves," took pen in hand to describe what it was to suppress the rebellion in the area about Charleston. The article reads as follows:

Camp Red House, Western Virginia,
Oct. 19, 1861.

The steamboat *Izetta*, carrying horses and army stores, was fired upon by rebel cavalry the forenoon of Oct. 11, at Winfield, Va. As soon as the intelligence was received by Col. Piatt at Camp Piatt, ten miles above Charleston, he ordered out 500 men under command of Lt.-Col. Toland with directions to proceed to Winfield, and there land the forces and pursue the rebels.

In one hour after the order was issued, 500 zouaves and all their arms and equipment were aboard the *Silver Lake* steamboat, making rapid headway down the Kanawha. They arrived at Charleston at midnight of the 11th and were delayed until 7 o'clock next morning. Col. Guthrie accompanied the force across the river to Red House and "captured" goods in a store belonging to a rebel (a small village grocery). Having failed to find any rebels (armed ones), the Zouaves decided to do the next best thing, which was, in their opinion, to get chickens for supper. The poor feathered tribe was doomed to a fearful end. More than a thousand of them were sacrificed to appease the stomachs made hungry by a fatiguing march. In less than an hour after our arrival soldiers might have been seen in every part of our company, munching chickens and brandishing chicken legs.

Scouting parties were sent out to scour the country; also foraging parties to take possession of such rebel property as would be useful to the government. Orders were issued against soldiers taking anything without orders from commanding officers.

On Monday Gen. Rosecrans ordered Col. Piatt to send his whole force (from Piatt, now Belle) and he arrived with the remaining companies of our regiment. We were ready to return when he arrived. However he concluded to see for himself what the country produces. We had already captured a large amount of property belonging to prominent secesh.

He ordered a delay and sent out more parties in search of rebel goods. The success which attended these parties shows, either the peculiar aptness of the Zouaves in capturing and confiscating secesh property, or the remarkable productiveness of the country in such goods.

We started back on Tuesday, Oct. 15th, having taken 75 head of cattle, 50 horses, 15 yoke of work oxen, 150 sheep, 30 barrels of flour, 2,000 hams, fine Virginia tobacco, and dry goods, notions from stores.

Five prominent Secessionists were taken prisoner and marched on foot with the company. The appearance of the regiment on the march in return, was novel and amusing in extreme—men, cattle, sheep, Zouaves mounted on mules and horses; wagons loaded with every variety of secesh valuables; the prisoners marching under guard, the whole forming a cavalcade not unlike the old Roman triumphal entries which attended Pompey and the Caesars in the days of regal pomp and pride. Our regiment came into camp in perfect order, though I imagine our Cincinnati friends would hardly have recognized us as the same body of men who passed through the city a few weeks ago on our departure for the field.

Colonel Toland and his officers fully appreciate the "principle that those who seek to destroy our government should not enjoy its protection." We are now stationed at Red House on the left bank of the river. We are drilling while detachments are searching the country for rebels. We promise you that the 34th (regiment) will not be behindhand in fighting or any other duty called on to perform.

Signed: Kappa.

And, the soldier correspondent might have added, a good time was had by all.

Union & Confederate Correspondence Relating to the Early 1861 Campaign in the Lower Kanawha Valley

■ CONFEDERATE

Putnam Court-House, April 21, 1861

His Excellency John Letcher,
Governor of Virginia:

Dear Sir: The people of this valley and adjacent counties are unarmed, with the exception of two companies at Kanawha Court-House and one in this county, and they have good reasons to apprehend that an organization is being formed in Ohio to enter this valley at the mouth of the Kanawha at Point Pleasant, for the purpose of robbing and murdering the people of Mason, Putnam, Kanawha, and other counties. We want at least eight or ten cannon on this river and arms sufficient to arm the whole people. A sufficient supply of arms for the people of Kanawha, Mason, Putnam, Logan, Boone, Nicholas, and Fayette ought to be sent at once to the Falls of Kanawha or some point lower down on the river. Give the people arms and they will rise en masse and defend themselves, and every county in this section will send one or more companies to defend the State or to fight wherever you may command them to go to fight for the cause of Virginia and the South. I believe the people in this section will sustain your proclamation and the action of the convention with great unanimity. Every hour the people are becoming more united, determined, and enthusiastic. All past differences are being forgiven, and the people swear to stand by each other and follow the flag of Virginia wherever it goes. Let us have arms as speedily as possible and the people will rise and fight. Arms could not be safely sent from Parkersburg to the mouth of the Kanawha at Point Pleasant. They would be seized in all probability at Pomeroy, Ohio.

I am, with great respect, yours truly,
J.G. NEWMAN

■ CONFEDERATE

Putnam Court-House, Va.,
April 22, 1861

His Excellency John Letcher,
Governor of Virginia:

Dear Sir: A gentleman of this county of much credibility, who has just returned from a trip through several counties in Ohio, says he was informed by several of his customers (he being a tobacconist) that efforts were now being made in several of the neighborhoods in Gallia and Jackson Counties, Ohio, to raise a sufficient force to invade this portion of Virginia, and produce an insurrection among the slaves and lay waste to the valley of the Kanawha. Believing this statement to be true, I am induced to write to you, and suggest the propriety of ordering one or more volunteer companies to Point Pleasant, the mouth of the Kanawha River. Buffalo is situated some twenty-two miles up the Kanawha River, but within some twelve miles of the Ohio River. As the Ohio River runs nearly parallel with the Kanawha from Point Pleasant to a point some eighteen or twenty miles below mouth of Kanawha, Buffalo would be the most accessible point to the abolitionists of Ohio to enter the valley of the Kanawha. There are more slaves in the neighborhood of Buffalo than there are from Buffalo to Point Pleasant. I would therefore also suggest the necessity of stationing some one or more companies at Buffalo. If we had arms we could soon raise a force to protect ourselves and give to other portions of the State the services of our volunteer companies. The people of this country are heart and soul with you in the defense of the State.

Very truly, your obedient servant,
ROBT. T. HARVEY

(Enclosure)

At a county court, held for the county of Putnam at the court-house thereof on Monday, the 22nd of April, 1861, it was ordered that the sum of $3,000 be appropriated for the purpose of purchasing arms of the people of said county, to be used for defending themselves and the State of Virginia, and that said sum be levied for and collected, and also the sum of $200 to pay the expenses of a special messenger to Richmond City to be levied and collected in the same manner. But this levy of said $3,000 is only to take effect in the event that the Governor of the Commonwealth of Virginia fails or refuses to supply said arms. And it is further ordered, that Maj. L. L. Bronaugh be, and he is hereby, appointed special commissioner to wait upon the Governor of Virginia for the purpose aforesaid.

Test:

ROBT. T. HARVEY
Clerk

■ UNION

Headquarters Ohio Volunteer Militia,
Columbus, Ohio, April 27, 1861.

Lieut. Gen. Winfield Scott,
Commanding U. S. Army:

. . . With the active army of operations it is proposed to cross the Ohio at or in the vicinity of Gallipolis and move up the valley of the Great Kanawha on Richmond. In combination with this Cumberland should be seized and a few thousand men left at Ironton or Gallipolis to cover the rear and right flank of the main column. The presence of this detachment and a prompt movement on Louisville or the heights opposite Cincinnati would effectively prevent any interference on the part of Kentucky. The movement on Richmond should be conducted with the utmost promptness, and could not fail to relieve Washington as well as to secure the destruction of the Southern Army, if aided by a decided advance on the eastern line. . . .

. . . I have the honor to be, general, very respectfully, yours,

GEO. B. McCLELLAN,
Major-General, Commanding Ohio Volunteers.

■ CONFEDERATE

Monday, April 29, 1861

The Governor submitted to the council for advice nominations of the following officers for the volunteer service of the State, viz: For lieutenant-colonels, William Mahone, of Norfolk City; *John McCausland, of Mason County*; Robert B. Chilton, late of the U. S. Army, and A. S. Taylor, late of the U. S. Marine Corps.

For the office of major: P. R. Page, of the county of Gloucester, J. P. Wilson, of Cumberland; Alonzo Loring, of Wheeling; Francis M. Boykin, Jr., of Lewis County, and Cornelius Boyle: Advised unanimously that the appointments be made. Ordered, that the commissions be issued accordingly.

JOHN LETCHER

■ CONFEDERATE

Headquarters Virginia Forces,
Richmond, Va., April 29, 1861.

Lieut. Col. John McCausland:

You will proceed to the valley of the Kanawha, and muster into the service of the State such volunteer companies (not exceeding ten) as may offer their services, in compliance with the call of the governor: take the command of them, and direct the military operations for the protection of that section of country. Your policy will be strictly defensive, and you will endeavor to give quiet and assurance to the inhabitants. It has been reported that two companies are already found in Kanawha County, Captain Patton's and Captain Swann's, and that there are two in Putnam County, Captain Becket's and Captain Fife's. It is supposed that others will offer their services. The number of enlisted men to a company, fixed by the Convention, is eighty-two. You will report the condition of the arms, etc., of each company, and, to enable you to supply deficiencies, five hundred muskets, of the old pattern, will be sent. I regret to state that they are the only kind at present for issue. Four field pieces will also be sent you as soon as possible, for the service of which you are desired to organize a company of artillery. The position of the companies at present is left to your judgment, and you are desired to report what points below Charleston will most effectually accomplish the objects in view.

I am, sir, etc.,

R. E. LEE,
Major-General, Commanding

Headquarters Virginia Forces
Richmond, Va., May 3, 1861.

Col. C. Q. Tompkins,
Charleston, Kanawha County, Va.:

Colonel: I have the honor to inform you that you have been appointed colonel of Virginia volunteers. Your commission is herewith forwarded to you. If you accept, you will take command of such troops as may be called out in Kanawha under the proclamation of the governor.

Lieut. Col. John McCausland has been previously directed to muster into the service such companies as may volunteer under the call of the governor. You will take measures to secure the safety and quiet of that county. Report what point you will occupy for the purpose.

Four field-pieces, 6-pounders, and some muskets, have been sent to the Kanawha Valley, subject to the order of Lieut. Col. John McCausland.

I am, sir, very respectfully,

R. S. GARNETT,
Adjutant General.

(Garnett was killed a month later.)

■ UNION

Headquarters Department of the Ohio,
Cincinnati, May 26, 1861.

To the Union Men of Western Virginia:

Virginians: The General Government has long enough endured the machinations of a few factious rebels in your midst. Armed traitors have in vain endeavored to deter you from expressing your loyalty at the polls. Having failed in this infamous attempt to deprive you of the exercist of your dearest rights, they now seek to inaugurate a reign of terror, and thus force you to yield to their schemes, and submit to the yoke of the traitorous conspiracy dignified by the name of Southern Confederacy.

They are destroying the property of citizens of your State and ruining your magnificent railways. The General Government has heretofore carefully abstained from sending troops across the Ohio, or even from posting them along its banks, although frequently urged by many of your prominent citizens to do so. I determined to await the result of the late election, desirous that no one might be able to say that the slightest effort had been made from this side to influence the free expression of your opinion, although the many agencies brought to bear upon you by the rebels were well known.

You have now shown, under the most adverse-circumstances, that the great mass of the people of Western Virginia are true and loyal to that beneficent Government under which we and our fathers have lived so long. As soon as the result of the election was known the traitors commenced their work of destruction. The General Government cannot close its ears to the demand you have made for assistance. I have ordered troops to cross the river: They come as your friends and brothers—as enemies only to the armed rebels who are preying upon you. Your homes, your families, and your property are safe under our protection. All your rights shall be religiously respected.

Notwithstanding all that has been said by the traitors to induce you to believe that our advent among you will be signalized by interference with your slaves, understand one thing clearly—not only will we abstain from all such interference, but we will, on the contrary, with an iron hand, crush any attempt at insurrection on their part. Now that we are in your midst, I call upon you to fly to arms and support the General Government.

Sever the connection that binds you to traitors. Proclaim to the world that the faith and loyalty so long boasted by the Old Dominion are still preserved in Western Virginia, and that you remain true to the Stars and Stripes.

GEO. B. McCLELLAN,
Major-General, Commanding.

(inclosure No. 6)
Address to the Soldiers of the Expedition,

Headquarters Department of the Ohio
Cincinnati, May 26, 1861.

Soldiers: You are ordered to cross the frontier and enter upon the soil of Virginia. Your mission is to restore peace and confidence, to protect the majesty of the law, and to rescue our brethren from the grasp of armed traitors.

You are to act in concert with the Virginia troops, and to support their advance. I place under the safeguard of your honor the persons and property of the Virginians. I know that you will respect their feelings and all their rights. Preserve the strictest discipline; remember that each one of you holds in his keeping the honor of Ohio and of the Union.

If you are called upon to overcome armed opposition, I know that your courage is equal to the task; but remember that your only foes are the armed traitors, and show mercy even to them when they are in your power, for many of them are misguided. When under your protection the loyal men of Western Virginia have been enabled to organize and arm, they can protect themselves, and you can then return to your homes with the proud satisfaction of having preserved a gallant people from destruction.

GEO. B. McCLELLAN,
Major-General, U.S. Army,
Commanding Department.

■ CONFEDERATE

Falls of Kanawha, Va., May 27, 1861.

Adjutant General Garnett, Richmond, Va.:

Sir: I have this moment an express from Lieutenant-Colonel McCausland, at Buffalo, dated yesterday, stating, "The Government has sent two hundred men to Gallipolis, and will have six hundred more here today. We are informed that they are intended to attack this camp. Send down all the troops you have." In addition to this, reliable information reaches me that large numbers of troops are concentrating at Oak Hill, twenty-three miles back of Gallipolis, and also at other places along the border. My idea is that these troops have been thrown into this proximity in order to overawe the loyal citizens of that region. For further particulars I beg leave to refer you to the bearer of this, Mr. David Kirkpatrick, a resident of this valley, and a well informed man.

Very respectfully,

C.Q. TOMPKINS,
Colonel, Virginia Volunteers, Commanding.

■ CONFEDERATE

Falls of Kanawha, Va., May 27, 1861

Col. R. S. Garnett, Adjutant-General:

Sir: I avail myself of a few moments' delay of the state to explain more fully the nature of my communication this morning. I consider it of sufficient importance for the employment of a special messenger, and accordingly have instructed the bearer, Mr. Kirkpatrick to convey this in person, and to telegraph from Staunton its import. I shall of course proceed to Buffalo as rapidly as possible. The idea is that the enemy intend crossing the Ohio River, to attack the camps at Buffalo. Unless they come in greatly superior force, we shall drive them back. On the other hand, if his numbers are large and the disaffection of the inhabitants strongly evinced, I shall take the most defensible position I may find, and rally the volunteers now in process of formation in the adjoining counties. Great excitement prevails in this region. The divided sentiment of the people adds to the confusion, and, except the few loyal companies now mustered into the service of the State, there are few of the people who sympathize with the secession policy. I send a special messenger (Mr. Kirkpatrick), because he is familiar with the whole of Ohio border, and can give you valuable information as to the resources, distances, & c. Mr. Kirkpatrick is reliable and intelligent. It is very desirable that Mr. Kirkpatrick should be the purveyor of some supplies for the troops which cannot be procured here. I beg that the quartermaster may be instructed to forward by him material for tents, three hundred blankets, five hundred cartridge-boxes (musket), and ten thousand percussion caps (rifle), &c.

In great haste, yours, respectfully,

C.Q. TOMPKINS,
Colonel, Virginia Volunteers, Commanding.

Headquarters Department of the Ohio,
Cincinnati, May 30, 1861

I am happy to say that the movement has caused a very great increase of the Union feeling. I am now organizing a movement on the valley of the Great Kanawha; will go there in person, and endeavor to capture the occupants of the secession camp at Buffalo, then occupy the Gauley Bridge, and return in time to direct such movements on Kentucky and Tennessee as may become necessary.

I will make a more detailed report when I receive Colonel Kelley's full report.

I am, very respectfully, your obedient servant,
GEO. B. McCLELLAN,
Major-General, U. S. Army
Lieut. Col. E. D. Townsend, A. A. G.

■ CONFEDERATE

Col. R. S. Garnett

Kanawha Court House, Va., May 30, 1861

Sir: The threatening aspect of affairs in this quarter induces me to send Lieutenant-Colonel McCausland to explain in detail matters that could not be discussed by letter. He will inform you of the disaffection of this population and the difficulty of procuring reliable troops for the emergency. There can be no doubt now that it is the intention of the enemy to occupy as much of this country as he may find open to invasion, and your attention is specially called to the necessity of sending, as early as practicable, a force at least sufficient to hold this valley in security. I have now under my command here three hundred and forty men, and when the companies now in process of formation in this valley shall have been completed it is probable their numbers will not exceed one thousand men. It is doubtful, in my mind, whether the militia will obey a call to the field. For these reasons it would seem proper that re-enforcements should be sent from such sources as you may deem proper. I beg leave, respectfully, to urge the importance of sending rifles, with suitable ammunition and I again request that staff officers for this department may be drawn from the troops comprising this command.

Very respectfully,
C.Q. TOMPKINS
Colonel, Virginia Volunteers, Commanding

■ CONFEDERATE

His Excellency Governor Letcher

Charleston, Kanawha County, Va., May 30, 1861

Men of Virginia! Men of Kanawha! To Arms!

The enemy has invaded your soil and threatens to overrun your country under the pretext of protection. You cannot serve two masters. You have not the right to repudiate allegiance to your own State. Be not seduced by his sophistry or intimidated by his threats. Rise and strike for your fireside and altars. Repel the aggressors and preserve your honor and your rights. Rally in every neighborhood with or without arms. Organize and unite with the sons of the soil to defend it. Report yourselves without delay to those nearest to you in military position. Come to the aid of your fathers, brothers and comrades in arms at this place, who are here for the protection of your mothers, wives, and sisters. Let every man who would uphold his rights turn out with such arms as he may get and drive the invader back.

C.Q. TOMPKINS,
Colonel, Virginia Volunteers, Commanding

Typical terrain Civil War soldiers fought over in southern West Virginia.

The Battle of Winfield

In October 1864, Federal Capt. John M. Reynolds, commanding Co. D, 7th W. Va. Cavalry, was sent to occupy Winfield, the county seat of Putnam County. With the 80 men under his command he was ordered to establish a fortified position overlooking the steamboat chute through Red House Shoals. Their mission was to stop all Confederate river traffic through the shoals.

In a short time they had completed a well-constructed installation of rifle pits along the river front. The breastworks not only protected them from the river side, they could also cover any threat from the rear.

Col. Vincent "Clawhammer" Witcher was operating with his 34th Virginia Cavalry on Mud River when he heard that the Federals had taken over Winfield. He reacted at once by sending two companies of Thurmond's Partisan Rangers to drive them out. The Rangers were led by their veteran leader and founder, Capt. Philip J. Thurmond of Monroe County. They were a hard-riding partisan troop that were usually not attached to any regular Confederate army unit. They basically were organized to protect their home neighborhood, but they soon found they could contribute more to the cause by collaborating with various established Rebel units.

About 400 rangers reached Winfield early in the morning of Oct. 26 and launched an attack through the streets of the town around 3:00 a.m. An erratic fire fight was kept up in the dark for over an hour but failed to dislodge the Federals from their strong position. Near 4:00 a.m., Captain Thurmond was mortally wounded in front of the courthouse when he was shot through the abdomen. By then it was apparent that the attack was a failure and the Confederates began to withdraw.

Since Captain Thurmond could not be moved, his younger brother Elias allowed himself to be captured so he could stay with his brother until he died. Later that morning Lt. William Bahlmann appeared in front of the Federal pickets under a flag of truce. He had with him a coffin and asked permission to come through the lines and bury Capt. Thurmond. Permission was granted and Capt. Philip J. Thurmond was buried in an unmarked grave thought to be behind the Hogue house, which is still standing near the courthouse.

The pro-union Charleston newspaper, *The West Virginia Journal*, reported: "Our men took five prisoners, among them Elias Thurmond, who is here in the guard house and wounded seven of the Rebels besides killing Phil Thurmond. Our loss was two slightly wounded."

As for the citizens of Winfield remembering much about the skirmish, they saw little of the action unless they dared look out their windows in the dark. The firing only lasted an hour or so and by daylight the Confederates were gone. Actually, the death of Captain Thurmond was about the only noteworthy result of the action at Winfield. Not much else was gained or lost on either side.

The small town of Winfield, county seat of Putnam, is seen here from the north side of the river. The valley has begun to widen out and the river is broad as it wends its way to its rendezvous with the Ohio.

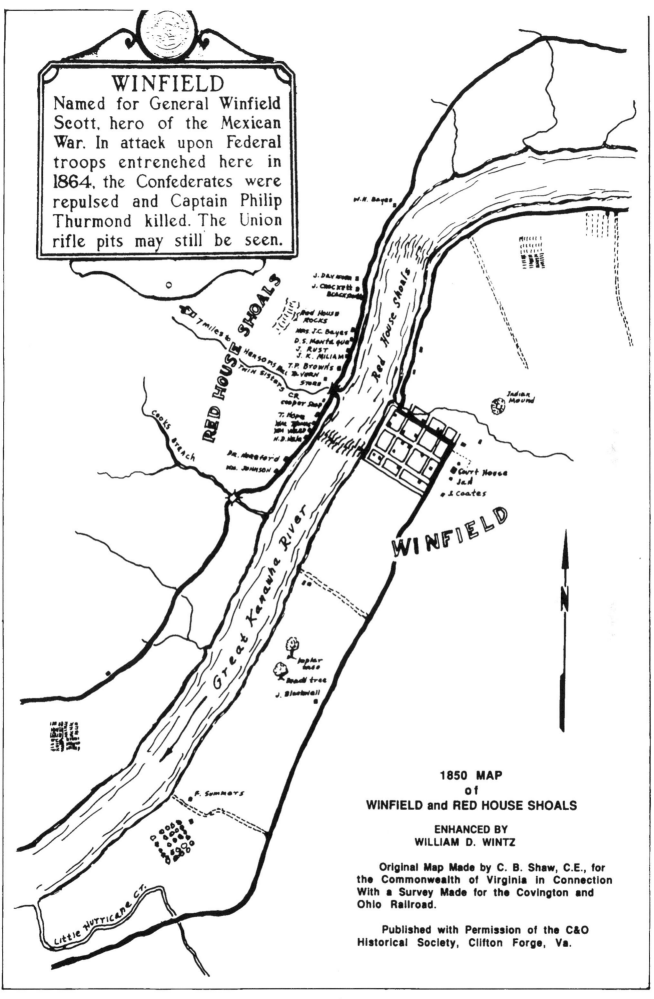

WINFIELD
Named for General Winfield Scott, hero of the Mexican War. In attack upon Federal troops entrenched here in 1864, the Confederates were repulsed and Captain Philip Thurmond killed. The Union rifle pits may still be seen.

1850 MAP
of
WINFIELD and RED HOUSE SHOALS

ENHANCED BY
WILLIAM D. WINTZ

Original Map Made by C. B. Shaw, C.E., for the Commonwealth of Virginia in Connection With a Survey Made for the Covington and Ohio Railroad.

Published with Permission of the C&O Historical Society, Clifton Forge, Va.

A Tale of Two Doctors

Just prior to the Civil War two doctors came to Poca Bottoms on the Kanawha River. They moved their families on adjoining farms located on the present site of the John Amos Power Plant. They were Dr. John J. Thompson and his son-in-law Andrew R. Barbee.

Andrew Russell Barbee was the son of Andrew R. and Nancy (Britton) Barbee and was born in Hawsbury, Rappahannock County, Va. In 1852 he married Margaret A. G. Thompson, daugther of Dr. John J. and Ann R. (Arthur) Thompson.

Dr. Barbee was educated at Petersburg, Va., where he studied medicine under Dr. John J. Thompson of Luray, Va. He also attended lectures at the University of Pennsylvania in 1848–49, and Richmond Medical College in 1849–50. He graduated as an M.D. from the University of Pennsylvania in 1851. He first located at Flint Hill, Va., and afterwards removed to Madison County, Va.

Why had the two doctors left their well-established practices in Old Virginia and moved to Poca Bottoms? There was not enough medical business for one physician in the area, not to mention two. Although they had soon established a small practice between them, before long it became apparent that their main labors were directed toward farming.

The real reason, however, that had compelled them to leave their well-developed home land and move to a strange place like Poca Bottoms had nothing to do with their professions or new pursuits. It was soon discovered by their new neighbors that both Dr. Thompson and Dr. Barbee were strongly set against separation from the Union. Therefore, probably the real reason they had come was to remove from a hotbed of secessionalism and relocate in a border area where their unpopular convictions might better be tolerated by the local citizens.

When war clouds began to gather, however, both men were forced to reconsider their political principles when it came time to choose sides. At the outbreak of the war Dr. John Thompson declared neutrality, even though his 18-year-old son was called home from VMI to help train Confederate soldiers at nearby Camp Tompkins.

Dr. Barbee, however, when faced with the same decision, was compelled to remain loyal to Virginia. Once he had cast his lot with the Confederacy, he immediately set about raising an infantry company that came to be known as the Border Riflemen. After the Battle of Scary, this unit became Co. A, of the 22nd Virginia Infantry.

Captain Barbee's company was quick to gain recognition when they formed a strong picket line that first repulsed the Federal troops at Scary Creek. Later Captain Barbee was promoted to colonel and it was said he would lead his men in combat all day and spend most of the night attending to their wounds. In 1863, at the Battle of White Sulphur, he was also wounded. After his recovery, Colonel Barbee was assigned to General Breckenridge's staff and when the general was called to serve in the Confederate government, Barbee temporarily became commander of the Department of the Southwest Virginia.

After the war, Dr. Barbee moved his practice to Buffalo, and in 1868 he relocated to Point Pleasant. In 1881 he was elected to the West Virginia Senate and was instrumental in securing passage of the law that regulates the practice of surgery and medicine in the state. He was also president of the West Virginia Medical Association and president of the local school board.

Dr. John Thompson, on the other hand, during the first part of the war, continued secretly to support any movement that might shorten or call off the conflict although he had claimed neutrality. According to a letter dated June 9, 1861, from Capt. George S. Patton to his commanding officer Col. C. Q. Tompkins, he reported the following concerning Dr. Thompson: "I am credibly informed that Dr. John Thompson, a member of the Virginia Legislature, is engaged in circulating a petition to the Governor of Virginia asking him not to send any more troops to the valley and to disband those already raised. This information came through 1st Sgt. Dudding of Capt. Barbee's company who is entirely reliable."

Gen. H. A. Wise's written orders dated July 19th to Colonel Tompkins, after the Battle of Scary, also referred to Dr. Thompson as follows: "Send the prisoner Dr. Thompson, whom I wish to deal with, strictly to me, at this place."

One other item referring to Dr. Thompson's activities was also included in the letter from Captain Patton to Colonel Tompkins, which stated: "Another circumstance of suspicious character, is the fact that within the past week a large supply of coffee and sugar has been landed at Winfield—a much larger supply than was ever known to be brought to that point. Arriving there at the same time were Miller, Waggoner, and Boreman, which renders the matter doubly suspicious."

Another story in connection with the above incident was told to the writer by the venerable Capt. Sid Morgan, whose family lived near the Thompson farm.

When Dr. Thompson's son, Cadet John K. Thompson, returned from VMI to train Confederate recruits at Camp Tompkins, he naturally came home first. Arriving after dark, he was met at the gate by one of the slaves. He was informed that he would have to sleep in the barn that night as his father had visitors who included Arthur I. Boreman and several other gentlemen of questionable loyalties.

Cooney Ricketts the Boy Soldier

Gen. Albert G. Jenkins wrote about Ricketts as follows: "When we were in camp at Greenbottom, this boy, who is now only 14 years old, had been drilling with us whenever he could borrow a horse. When we were ordered to Camp Tompkins he went to his uncle's pasture and appropriated this frisky little mule and began following the company. When I was told he was behind us I sent him word to return home, but he only dropped back some and continued tagging along. Finding my orders disregarded, I rode to the rear of the company and commanded him to return as I knew his mother, who was a widow, would be nearly deranged, especially since we already had in our company one of her other sons, Albert Gallatin Ricketts, who

was a namesake. However, towards night my men informed me that the boy had continued to follow the company all day. This time when I went back to reason with him he said, 'Captain, the road is free, I will ride in sight of you by day and camp nearby at night.' When I told him he was too young for duty he said, 'I can carry water, I can wait on you, and I can do anything else, but I am determined to go to war.' We could hold out no longer and that night a vote was taken in our company and we adopted him as 'The Child of the Regiment.'"

He always went by the name "Cooney" Ricketts and he usually rode at the head of the Rebel column beside his beloved captain, who was also evidently very fond of him. When General Wise came to visit Camp Tompkins, Cooney was introduced to the General as "The Child of the Regiment."

I will mention a circumstance that occurred while Captain Jenkins' company was camped directly opposite the brick house just below Tackett's Creek where I lived. They were in great need for pistols as they could no longer receive any from Cincinnati. The captain conceived a plan to obtain some which involved the youthful, but brave Ricketts. Furnished with an ample supply of money, he was started to Cincinnati astride his little mule. He rode to Guyandotte where he hid his mule with relatives and boarded a steamer for Cincinnati. Arriving in the city, so as not to be suspected, he began buying up pistols here and there until he obtained the prescribed number. He then went out to the Union Camp Dennison where he collected important information in regard to troop numbers and movement.

After his successful return, Cooney was nicely equipped with a bridle and saddle for his mule. As he and the mule were small, the ladies all thought him cute, sweet, and brave. A great many tried to persuade him to return home, but all in vain.

Unofficial records indicate that Cooney Ricketts continued to ride with the 8th Virginia Cavalry until his 16th birthday. Since he was under age, he was never officially assigned or listed on any unit rosters. On Aug. 22, 1863, his 16th birthday, Lucian "Cooney" Ricketts was enrolled at the Virginia Military Institute at Lexington, Va.

At the battle of New Market he was a cadet private in Co. C. Col. Francis L. Smith wrote on Feb. 8, 1910, about the battle: "I would like to see some mention of 'Cooney' Ricketts for I thought he acted with great gallantry, mounted on Shaw's horse, riding ahead and in front of the corps when we became engaged with the enemy."

Ricketts died in 1906 and is buried in Huntington's Spring Hill Cemetery.

Gallipolis, Ohio

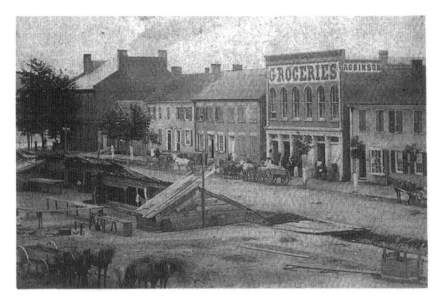

This Civil War photo of the public square in Gallipolis shows us what the Union bastion looked like 1861–65.

This 1995 view of the above scene is fascinating as several of the Civil War-era buildings still stand. Note the twin chimneys on the left. Riverboat landing is next to this block.

Gallipolis is a charming and historic little city. This view of the city boat landing is on the Ohio River where most of the Federal troops must have embarked. The building with many windows is the same as pictured above.

This scenic photo of Gallipolis, the Ohio River and Point Pleasant in the distance was taken from a commanding height known as "fortification hill." Obviously every town that was near a hill had a fort to cover the river.

A quiet little piece of Civil War history reposes on Pine Street in Gallipolis, Ohio. This Civil War soldiers' Federal cemetery contains over a hundred Union graves.

Although we cannot say for certain, this Union soldier probably died of wounds received in the days leading up to the Battle of Charleston. His grave is in the Pine Street Federal cemetery, Gallipolis, Ohio.

During Wise's Retreat, Victoria Hansford wrote: "I remember seeing five or six fine looking officers riding by on beautiful horses. They made a splended apperance in their dark gray uniforms with brass buttons. They wore plumes in their hats witch they carried across there arms until they had passed our group of weeping waving women. One in particular I noticed was a tall handsome man riding a beautiful black horse. Inquiring who they were I was told they were officers of Chapmans Battery from Monroe County." Victoria failed to mention that the officer she saw that day was Lt. Thomas Teays and after the war they were married at her home in Coalsmouth now St. Albans.

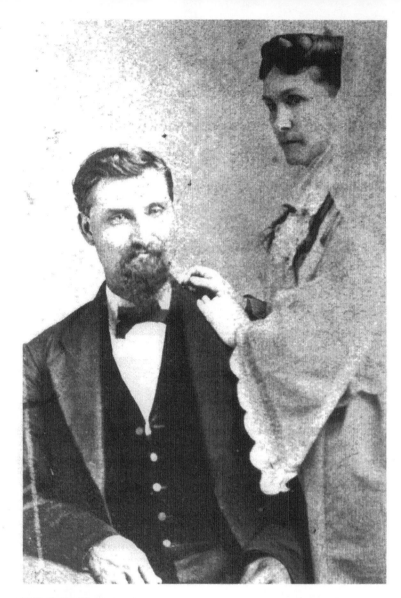

Teays Tavern was built in 1831 by James T. Teays. It was a large 12-room, 2-story frame house that stood on the lower side of Coal River on the James River and Kanawha River turnpike. In 1852 James Teays and Col. Philip Thompson built a covered bridge at that place and the tavern became a stop for the four-horse stage. It was used as a hotel until 1876, when it became the residence of the Teays and Barker families until the 1930s. The property was sold and it was torn down circa 1950.

Blue Tide
Military Occupation, 1863–65

War in the Kanawha Valley, 1863

Mr. J. C. McFarland, president of the Branch Bank of Virginia and a very prominent citizen and businessman in Charleston, wrote to Indiana Hornbook of Wheeling on Jan. 5, 1863, concerning wartime conditions in Charleston. He noted that the retreating Federals had "set fire to their warehouses and the Bank and the Kanawha House [both of which he owned or controlled] and others [buildings] in the center of town. By the coming of the Confederates, two things were accomplished, viz the getting of several thousand barrels of salt and the marriage of two young girls in town." Another letter of March 17, 1863, two days after Col. Hayes and the 23rd Ohio arrived at Camp White, still noted the troubled times in the City. Mr. McFarland said that not a fence was standing, there were 100 teams in the streets and sidewalks were ruined. The Branch Bank of Virginia and Kanawha House had occupied an entire block but the fire-damaged walls had been blown down by gusty winds leaving a huge mass of rubble. The only building still intact, he wrote, was the Presbyterian Church but its lecture room had been converted into a kitchen for teamsters and "our village presents a most forlorn and desolate appearance." It was to this scene that Col. Hayes led his occupation force.

By the time Col. Rutherford B. Hayes and his Twenty-third Ohio Volunteers reached Charleston in March 1863, they were seasoned veterans of several engagements. They had been part of General Rosecrans' army at the Battle of Carnifex Ferry, Sept. 10, 1861. William McKinley said of Carnifex Ferry:

This was our first real fight and the effect of the victory was of far more consequence to us than the battle itself. It gave us confidence in ourselves and faith in our commander. We learned that we could fight and whip the rebels on their own ground.

The regiment was involved in a fight at Princeton in April of 1862 and was driven to Flat Top Mountain, enduring severe hardships and near starvation. They joined McClellan's army in Washington later in the summer and were important combatants in the Federal victories at Frederick, Maryland, and South Mountain. They were also participants in the great Battle of Antietam Creek near Sharpsburg, Maryland, on Sept. 17, 1862.

Colonel Hayes was severely wounded in the left arm at South Mountain and was sent back to Ohio to recover. During Hayes' absence from the field, Sgt. William McKinley distinguished himself at Antietam to the extent that he was elevated to the rank of second lieutenant. Colonel Hayes noted in his diary on Dec. 13, 1862: "Our new second lieutenant, McKinley, returned today—an exceedingly bright, intelligent, and gentlemanly young officer. He promises to be one of the best." McKinley was returning to camp from a recruiting trip to Ohio.

Colonel Hayes had rejoined his regiment at Camp Maskell on Nov. 30, 1862, his wound well healed, but his arm not yet fully restored to normal strength. Camp Maskell was located on the south side of the Kanawha River, two miles from Gauley Bridge, in sight of the falls. Early in 1863 the name of the Camp was

Wedding portrait, Rutherford and Lucy Webb Hayes,
December 31, 1852. RBHI.

"Last Monday, the 15th, Lucy [his wife],
Mother Webb [his mother-in-law], and all
the boys came here from Cincinnati on the
'Market Boy'. A few happy days, when little
Joseph [his 18-month-old son] sickened and
died yesterday at noon (12:40). Poor little
darling! A sweet, bright boy, 'looked like his
father', but with large, handsome blue eyes
much like Webb's. Teething, dysentery, and
brain affected, the diseases. He died without
suffering; lay on the table in our room in the
Quarrier cottage surrounded by white roses
and buds all afternoon, and was sent to
Cincinnati in the care of Corporal Schirmes,
Company K [D], this morning."

COL. RUTHERFORD B. HAYES
diary entry on June 25, 1863 at Camp White, opposite Charleston

changed to Camp Reynolds, in honor of Maj. Eugene
E. Reynolds, who was killed at South Mountain.

The first two months of 1863 were spent in building log cabins at Camp Reynolds near Gauley Bridge.
Colonel Hayes said he had a double cabin with two
rooms, 18 by 20 each, and a breezeway measuring
6 by 18, between the rooms, all under a single roof.
He had stone fireplaces and chimneys. Work details
had built cabins for the men, dug drainage ditches,
made walks, roads and bridges.

Mrs. Hayes and two sons visited Camp Reynolds
from January 24 to Saturday, March 21, after the regiment moved to Charleston. Col. Hayes wrote in his
diary that the family "rowed skiffs, fished, built dams,
sailed little ships, played cards, and enjoyed camp life
generally."

On March 22, 1863, Hayes wrote that it seemed
they were intended for a permanent garrison at Camp
White (named for Col. Carr White of the Twelfth
Ohio) across from the mouth of Elk.

Rutherford B. Hayes was born in Ohio in 1822. When the war began, the Literary Club of Cincinnati formed a military drilling company with Hayes elected captain. He was later appointed a major of the 23rd Ohio Volunteer Infantry. He distinguished himself in several battles and earned rapid promotions. He resigned from the army on June 8, 1865, with the rank of major general. Hayes spent considerable time in the southern West Virginia campaign, including occupation duty in the Kanawha Valley. After the war he served in the U.S. Congress, was governor of Ohio and was elected President of the United States by a margin of one vote in the Electoral College in 1877. He served one term and died at his home, Spiegel Grove, in Fremont, Ohio, in 1893. RBHL.

Hayes and McKinley of the 23rd

In 1861 Rutherford B. Hayes and William McKinley, both sons of Ohio, donned the blue uniform of the United States and joined in a war to crush rebellion in Old Virginia.

Charleston has the historic distinction of being the headquarters of the 23rd Ohio Volunteer Infantry Regiment in which both Hayes and McKinley served.

Their faces were certainly familiar to the native Charlestonians and this writer's (Richard Andre) great-grandmother, Mrs. Sarah Robinson, who liked to recall serving supper to the two future presidents at her boarding house on Virginia Street. Mrs. Robinson's brothers were at that time on active duty with the Confederate Army.

It is interesting to note that Virginia has produced more American presidents than any other state with eight, followed by Ohio, which gave birth to seven.

Certainly more than a few of the gray-clad Confederates were blood kin and descendants of the Virginia presidents.

It is astonishing to consider that five of the seven Ohio presidents were Civil War veterans of the Union army. The first, U. S. Grant was elected in 1868, while all but one of the Virginia presidents held office *before* the Civil War.

The clash of the blue and gray armies in the Kanawha Valley—as across all the battlefields of the Civil War—was the bloody result of the conflict between the old and the new America. Hayes, the Cincinnati lawyer, and McKinley, the ironworker's son, were in many ways vanguards of the industrial revolution. While, on the other side, Lee and his men fought to uphold an agrarian South that was being left behind by the relentless efficiency of the machine age.

In a final ironic footnote to the history of Hayes and McKinley in the Kanawha Valley—the assassin who killed President McKinley in 1901, Leon Czolgosz, once worked in a nail factory in Charleston.

We shall probably be visited by the Rebels while here. Our force is small but will perhaps do. My command is Twenty-third Ohio, Fifth and Thirteenth Virginia, three companies of cavalry, and a fine battery. I have some of the best, and I suspect some of about the poorest troops in service. They are scattered from Gauley to the mouth of [Big] Sandy on the Kentucky line. They are well posted to keep down bushwhacking and the like, but would be of small account against an invading force.

Colonel Hayes consistently talked about the possibility of an attack of Charleston, and actually welcomed action! "Had a dispatch from Captain Simmonds at Gauley; he reports rumors of an early advance on all our posts. 'Sensational!' General Scammon in a 'stew' about it."

There were rumors of enemy movements in Boone and Logan counties, also on the Big Sandy, and Colonel Hayes thought the reported action was directed toward taking the Kanawha Valley and the salt works. His diary records the battle of Hurricane Bridge on March 28 and Jenkins' attack on Point Pleasant on March 29. Lieutenant Colonel Comly and five companies of the Twenty-third were dispatched by steamboat to Coalsmouth (St. Albans) to set up a defense.

"This is a beautiful valley from Piatt [Belle] down to its mouth. Make west Virginia a free State and Charleston ought to be a sort of Pittsburgh."

COL. RUTHERFORD B. HAYES
Monday, March 23, 1863

The Jenkins raid failed to penetrate farther than Hurricane and, except for causing some excitement, Colonel Hayes considered the raid to be a failure for the Confederate cause. "Yes, Jenkins made a dash into Point Pleasant, but he dashed out before doing much mischief with a loss of seventy-five killed and prisoners. He attacked one other post garrisoned by men under my command but was repulsed. His raid was a failure. He lost about one hundred and fifty men while in this region and accomplished nothing."

War was not all that the soldiers occupied themselves with. The pleasures of camp life along the Kanawha during the spring of 1863 were described by Colonel Hayes:

. . . Drilling, boating, ball playing make the time pass pleasantly.

You can stay on the opposite side of the river [Charleston] for seven dollars a week or in a comfortable tent on this side with better grub for nothing.

On July 7, after the building of Fort Scammon, Colonel Hayes in his diary: "Heard the news of Vicksburg captured. Fired one hundred guns and had a good time." It is presumed this was the second firing of the battery at Fort Scammon. On the sixth of July he had written to Mrs. Hayes:

We had a good Fourth. Salutes from Simmonds and Austin [artillery commanders]. A good deal of drinking but no harm. We let all out of the guard houses.

Colonel Hayes' command left Charleston for two weeks in July for a campaign in Raleigh County on Piney Creek. The Confederate force retreated and Col. Hayes' men destroyed the Rebel fortifications. They were then dispatched to engage Confederate cavalry raider John H. Morgan at Pomeroy, Ohio, on the Ohio River. The Twenty-third was part of a force which included gunboats, local militia, and cavalry. Morgan was defeated, losing 800 men.

Gen. B. F. Kelley came to Charleston in October to review the local troops, an event that probably caused considerable excitement and occupied the soldiers and their officers for several days in preparations for his visit.

The Kanawha Valley was quiet through the remainder of 1863. Indeed, Charleston had seen nothing of the war at all during the year except for the firing of Fort Scammon's guns to celebrate both the Fourth of July and the victory of Grant at Vicksburg.

"Charleston was a fine town before the war, and had a very cultivated society. The war broke it up, but now the town is gaining again and will ultimately recover its former prosperity."

COL. RUTHERFORD B. HAYES
October 10, 1863

---

*From *Fort Scammon and A History Of The Civil War In Charleston And The Kanawha Valley, West Virginia*, Paul D. Marshall & Assoc. Inc., 1986

Letters written by Rutherford B. Hayes to his wife, Lucy.

Camp White, April 5, 1863

Dearest

The weather is good, our camp dry, and everybody happy. Joe has got a sail rigged on his large skiff and he enjoys sailing on the River. It is pleasant to be able to make use of those otherwise disagreeable spring winds to do our rowing.

Visited the hospital (it being Sunday) over in town this morning. It is clean, airy and cheerful looking. We have only a few there—mostly very old cases.

Comley heard a couple of ladies singing Secesh songs as if for his ear in a fine dwelling in town. Joe has got his revenge by obtaining an order to use three rooms for hospital patients. The announcement caused grief and dismay—they fear smallpox (a case has appeared). I think Joe repents his victory now.

Enclosed photographs, except Comley's, are all taken by a Co. B. man who is turning a number of honest pennies by the means. Charlie Smith, Birch will recollect as Capt Avery's orderly.

Five companies of the 23rd have had a hard race after Jenkins. They got his stragglers. Colonel Paxton and Gilmore are after him with their cavalry. General J. has had bad luck with this raid. He came in with 700 to 800 men. He will get off with 400 to 500, badly used up and nothing to pay for his losses. We lost half a dozen killed. They murdered one citizen of Point Pleasant, an old veteran of 1812, aged 84. They will run us out in a month or two I suspect, unless we are strengthened, or they weaken. General S. is prepared to destroy salt and salt works if he does have to leave.

I think of you and the boys oftener than ever. Love to 'em and oceans for yourself.

Affectionately ever
R.

P.S. I sent by express $350, in a package with $200, of Joe's. It ought to reach Mother Webb in a day or two after this letter. Write if it *doesn't* or *does*.

Charleston, Camp Elk, July 2, 1864

Dearest

Back again to this point last night. Camped opposite the lower end of Camp White on the broad level bottom in the angle between Elk and Kanawha. My headquarters on one of the pretty wooded hills near Judge Summers'.

Got your letter of 16th. All others gone around to Martinsburg. Will get them soon. Very much pleased to read about the boys and their good behaviour.

Dr. Joe went to Gallipolis with our wounded, expecting to visit you, but the rumors of an immediate movement brought him back. We now have a camp rumor that Crook is to command this Department. If so we shall stay here two or three weeks; otherwise, only a few days, probably.

You wrote one thoughtless sentence, complaining of Lincoln for failing to protect our unfortunate prisoners by retaliation. All a mistake, darling. All such things should be avoided as much as possible. We have done too much rather than too little. General Hunter turned Mrs. Governor Letcher and daughters out of their home at Lexington and on ten minutes' notice burned the beautiful place in retaliation for some bushwhackers' burning out Governor Pierpont [of West Virginia]. And I am glad to say that General Crook's division officers and men were all disgusted with it.

I have just got your letter of June 1. They will all get here sooner or later. The flag is a beautiful one. I see it floating now near the piers of the Elk River Bridge.

Three companies of the Twelfth under Major Carey are ordered to join the Twenty-third today—Lieutenants Otis, Hiltz and _____ command them, making the Twenty-third the strongest veteran regiment. Colonel White and the rest bid us good-bye today. What an excellent man he is. I never knew a better.

You use the phrase "brutal Rebels." Don't be cheated in that way. . . . but it is very idle to get up anxiety on account of any supposed peculiar cruelty on the part of Rebels. Keepers of prisons in Cincinnati, as well as in Danville, are hard-hearted and cruel.

Affectionately,
R.

William McKinley was born in Ohio in 1843. When the war began he was the first man to enlist as a volunteer from his hometown of Niles, Ohio. He became commissary sergeant in the 23rd Ohio Volunteer Infantry, commanded by Rutherford B. Hayes. McKinley carried food and coffee to his regiment during the Battle of Antietam. His bravery under fire earned him a commission as a second lieutenant. While with is regiment, he spent some time in the Kanawha Valley. At war's end he had attained the rank of brevet major. Like Hayes, McKinley served in the U.S. Congress, was governor of Ohio and was elected President of the United States in 1896. Ironically, he was killed by an assassin who worked in Kanawha City, just a few miles from where McKinley was stationed in 1863. He died on Sept. 14, 1901. PHC

The 23rd Ohio Volunteer Infantry in Charleston

Camp White, May 2, 1863

We are fortifying, partly to occupy time, partly to be safe.

Camp White, May 7, 1863

We are building a fort on the hill above our camp--a good position.

Camp White, May 17, 1863

We have nearly finished a tolerable fort, and have a gunboat. I have thirteen pieces of artillery.

Camp White, May 17, 1863

We are in no danger here. We have built a tolerably good fort which we can hold against superior forces perhaps a week or two or more. We have a gunboat which will be useful as long as the river is navigable. My whole brigade has been here. The most of it is good and the rest is improving.

Camp White, May 25, 1863

The Rebels don't make much progress towards getting us out. We are tolerably well fortified here and at Fayette. At the latter place they tried it, banging away three or four days and doing nothing.

Camp White, May 27, 1863

We are sufficiently fortified to keep our position against anything but greatly superior forces.

The quotes above, from the pen of Col. Rutherford B. Hayes, are in reference to the building of Fort Scammon, located at the summit of an area known today in Charleston, West Virginia, as Fort Hill.

Colonel Hayes was the commander of the Twenty-third Ohio Volunteers, the regiment credited with the construction of Fort Scammon. The regiment had moved into Camp White at Ferry Branch on the south bank of the Kanawha River, opposite the mouth of Elk River, on March 15, 1863. In addition to the Twenty-third, Colonel Hayes also had in his command "the Fifth and Thirtieth Virginia, three companies of cavalry, and a fine battery." His command stretched from the mouth of Gauley River to the Kentucky line at the mouth of the Big Sandy River. He noted in his letters that his force was well posted to counter the bushwhacking warfare common in western Virginia, but would not be very successful in resisting an invasion force.

The Twenty-third of Ohio, the regiment of Fort Scammon history, was mustered into service of the United States on June 11, 1861. Enlistments were for three years. The first colonel of the regiment was William S. Rosecrans, but before the regiment left its first headquarters at Camp Chase near Columbus, Ohio, Colonel Rosecrans was promoted to brigadier general in the regular army. He was replaced by Col. Eliakim Parker Scammon, for whom the Charleston fortification was named. Scammon was commander until October 1862. Thereafter, the regiment was led by Rutherford B. Hayes.

Although Col. William Rosecrans would not be the commander of the Twenty-third during its occupation of the Kanawha Valley, he would have been in familiar territory. He had been captain in the U.S. Army Corps of Engineers but resigned his commission in 1853 to enter the engineering consulting business. He subsequently became the project engineer and part-owner of the Coal River Navigation Company, and also owned a coal company. The project design utilized eight locks and dams to provide slack water for 35 miles of the Coal River from Coalsmouth (St. Albans) to Peytona, as well as Lock and Dam A on Little Coal to reach mines in its watershed.

Captain Rosecrans resigned as president of the company when construction began in 1855 and left the area to build a canal coal oil refinery in Cincinnati, Ohio. The refinery had just become productive when he left Cincinnati to rejoin the army. It is interesting to note that Thomas L. Broun, who succeeded Rosecrans as president of the Coal River Navigation Company, became a Major in the Confederate army. Mr. Broun was also the law partner of George S. Patton, grandfather of the famous World War II general.

By the time of Fort Scammon's construction, all threats of Confederate control of the valley were over but, of course, this fact was not known to Colonel Hayes. He remained vigilant due, in part, to rumors of Confederate troop movements nearby, and certainly because of the real but generally ineffective raid by Confederate General Jenkins' party of 400 to 700 men on Point Pleasant and Hurricane in March of 1863. His concern was simply stated in a letter to Mrs. Hayes on April 5: "They will run us out in a month or two, I suspect, unless we are strengthened, or they are weakened. General Scammon is prepared to destroy salt and salt works if he does not have leave."

---

*From *Fort Scammon and A History Of The Civil War In Charleston And The Kanawha Valley, West Virginia*, Paul D. Marshall & Assoc. Inc., 1986

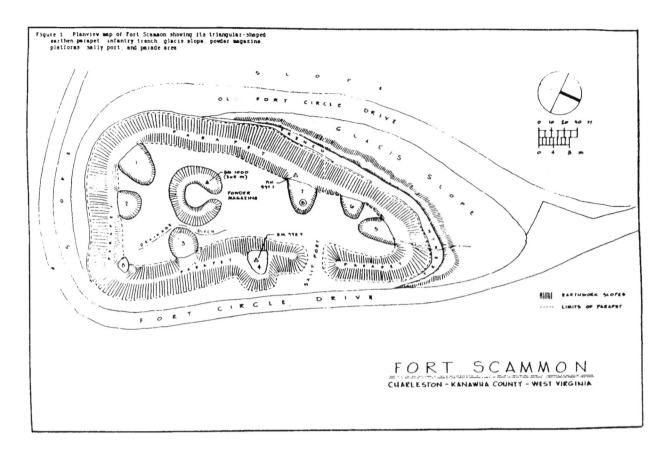

Figure 1 Planview map of Fort Scammon showing its triangular-shaped earthen parapet, infantry trench, glacis slope, powder magazine, platforms, sally port, and parade area.

FORT SCAMMON
CHARLESTON - KANAWHA COUNTY - WEST VIRGINIA

Fort Scammon

The construction of Fort Scammon was ordered by Col. Rutherford B. Hayes, commander of the Twenty-third. Part of Hayes' decision to build the fort atop Fort Hill was related to its high unobstructed view (1,000 ASL) of the Kanawha Valley and the obvious advantages this imparted to Union artillery. Building Fort Scammon on the hill served to strengthen the Union army's control over the strategic James River and Kanawha Turnpike, a road that followed the south side of the Kanawha River along roughly the same route as the present-day MacCorkle Avenue (Route 61).

Hayes' decisions about the specific location and construction, and hence defensibility, of Fort Scammon also were undoubtedly related to the steepness of its southeastern and northwestern approaches. The base of the hill on the southeastern, or Ferry Branch-Oakwood Road (Route 214) side, for example, is nearly 350 feet below the elevation of the fort. On the fort's northwestern side, the hill descends into a narrow, steep-sided hollow that is traversed by Danner Road. Southern and northern approaches to Fort Scammon have somewhat gentler slopes than the other approaches. One hundred feet or so below the south, or back, end of the fort, for instance, is a fair-

ly wide saddle that straddles (via Cantley Road) two hollows which branch off from the Ferry Branch and Danner hollows. Traveling from the north, or front, end of Fort Scammon toward the Kanawha River the grade changes from a steep slope to one characterized by two hundred feet of more level ground.

Hayes' decision to fortify Fort Hill was probably also related to the damage Confederate cannon inflicted on the Union army during the Battle of Charleston in 1862. During this battle, Confederate artillery on Fort Hill bombarded Union positions along the Elk River over a mile away (e.g. Watts Hill). Historical records, however, do not indicate from which part of Fort Hill these cannon were fired. Consequently, it is not known whether Confederate gun emplacements still exist on Fort Hill. At present, there is no evidence that indicates Confederate artillery was positioned at the site of Fort Scammon during or after the Battle of Charleston.

After Charleston was retaken in late 1862, the Union defenses on Fort Hill supposedly consisted of seven cannons at Fort Scammon and four cannons on the lower northern face of the hill. Part of this latter area, including a section of the road that came up the hill from Ferry Branch, was lopped off during the construction of Interstate 64. In addition to artillery positions, a camp for some or all of Fort Scammon's

- 138 -

troops was reportedly found in the saddle just south of the fort. Another camp located at the mouth of Ferry Branch, Camp White, was the main Union encampment in Charleston. Neither Camp White, the camp south of the fort, nor the artillery battery on the lower northeast face of Fort Hill, however, have ever been verified archaeologically.

While the location of Fort Scammon is an established fact, information about its configuration is lacking in the historical record. Maps showing the dimensions and structures of the fort, for example, are inaccurate or incomplete. Furthermore, very little has been written about the construction of the different structures that made up the site. Added to this lack of detail is a general lack of historical information about the weapons, ammunition, and other artifacts the Union soldiers used while manning the fort.

---

*From *Fort Scammon and A History Of The Civil War In Charleston And The Kanawha Valley, West Virginia*, Paul D. Marshall & Assoc. Inc., 1986

Modern views of the remains of Fort Scammon on Fort Hill, looking south. The site, owned by the City of Charleston, is on the National Register of Historic Places. It is approximately one acre in size and is the only Civil War fort located in Kanawha County. In 1986 an extensive survey and archaeological investigation of the powder magazine, outer trench and one of the seven artillery platforms was conducted by Paul D. Marshall & Assoc. of Charleston. Union army artifacts were excavated.

The Guerilla War

The guerillas, or bushwhackers, as they were more popularly known, were everywhere. They operated in a land of rugged mountain ridges, deep gorges and hollows, and countless thickets and glades. Just as important, their Confederate-sympathizing neighbors protected them, sustained them, and provided them with information concerning Union troop and supply movements. Thus, the bushwhackers were able to strike quickly and melt into the countryside, operating in relative security.

A correspondent of the *Cincinnati Times*, writing from Parkersburg in September 1861 said:

> . . . But though no rebel army may again appear this side of the mountains—and one is not likely to—the war in Western Virginia is far from being at an end. There is not a county in all this part of the Old Dominion that does not contain a greater or less number of Secessionists, who have degenerated into assassins. They are committing murders daily, lying in ambush for that purpose. Not only the Union volunteers, but their own neighbors . . . no neighborhood is safe from their depredations, unless protected by Federal bayonets. . . . They pillage, burn, destroy, and kill . . .

Guerilla warfare was officially adopted as a military policy by Gov. John Letcher, of Virginia, in a proclamation to the people of Virginia on March 10, 1862. In the proclamation he asked: "The loyal citizens of the West and North West, in counties not herein named, are earnestly invoked to form guerilla companies, and strike when least expected, once more for the state that gave them birth."

The proclamation was sanctioned by both the Virginia legislature and the Confederate Congress. Soon there were many bands of "partisans" organized to carry out Governor Letcher's policy. The guerilla bands in Western Virginia included the "Moccasin Rangers" (also called the Virginia State Rangers); the troops of Col. John Imboden; A. G. Jenkins Cavalry; Gen. John Morgan's Rangers; McNeill's Rangers; and others.

The Confederate high command, however, tried to discourage partisan warfare, because the bands gave refuge to deserters and were as likely to attack Confederate sympathizers as Union supporters. Thomas Rosser, a Confederate cavalryman, said: "they roam broadcast over the country, a band of thieves, stealing, pillaging, plundering . . . an injury to the cause." General Lee denounced them by stating: "I regard

the whole system as an unmixed evil." Nevertheless the rebel guerilla movement expanded as the Civil War dragged on.

The guerilla war was tough, dirty work for the Twenty-third Ohio, units of the 2nd, 3rd, 10th and 11th Virginia (Union) Infantry regiments, and the 6th Virginia Cavalry under Gen. Robert H. Milroy. Matters became worse in August of 1862 when troops of Gen. Jacob Cox were pulled out of the Kanawha Valley to aid Gen. John Pope east of the mountains. (The Kanawha forces were responsible for patrolling 25 counties.) At about the same time President Lincoln issued a call for 300,000 more troops. This created two problems. Fear of being drafted into the Union army caused some reluctant rebels to join guerilla bands, and the draft threatened to drain the ravaged West Virginia counties of able-bodied men needed to resist the bushwhackers at home.

The increase of guerilla activity and lawlessness through 1863 and 1864, including areas close to full-time Union occupation such as the Kanawha and Guyandotte valleys, caused considerable loss of confidence in the new Wheeling government. Loyal citizens had expected the new government to act positively to protect them and their property but the bushwhackers disrupted county and local governments, anarchy prevailed in many counties, and disaffection, neutralism, and secessionist sentiment were on the rise. During this period, then, the greatest impact of the guerilla forces was not military, but political.

Gov. A. I. Boreman, on Dec. 28, 1864, issued an address from Wheeling calling on West Virginia citizens to form vigilante groups and take justice into their own hands. His words reflected the bitterness and hate that had enveloped the new state, and he was also admitting that his government had failed to cope with the guerilla situation:

> For months past, bands of armed men have infested the State, stealing, robbing and murdering, and within a short time their numbers and the frequency of their outrages have alarmingly increased. Some of them claim to be rebel soldiers; but whatever they profess, they neither observe the rules of civilized warfare, nor regard the civilities of common thieves and robbers or the decencies of ordinary murderers. . . . It is evident that the presence of these predatory bands is known to the disloyal of the section infested by them, when unknown to others; and that those

persons do not disclose their presence to those who would be able to thwart their devilish purposes; but, on the contrary give these outlaws food and conceal-ment, with full knowledge of their purposes and practices. On account of the great numbers of these bandits, and the fact that they are always armed, it is impossible for the civil officers to arrest or the civil courts to punish them. . . . In view of this anomalous and alarming state of things, I earnestly recommend the loyal people of the State to organize themselves into Companies of such numbers as may be prac-ticable and expedient for the hunting down and cap-turing or killing these outlaws wherever they may be found—executing summary justice where found in the act or where they cannot otherwise be cap-tured—and thus aid the authorities of the State in restoring and preserving peace, order and security.

Interesting but confusing orders issued to Colonel Hayes by Gen. E. P. Scammon in May of 1863, just after completion of Fort Scammon:

HEADQUARTERS
Charleston, May 12, 1863

Col. R. B. HAYES
Commanding Brigade Opposite Charleston:
Your note is just received and all is right. [No record has been found as to content of Hayes' notes.] There is, of course, great commotion in town. All will think it horrible, but they have been playing the game of treachery and must take the result. I have ordered 100 men to be brought over as a firing party. [Was this to be a detail for torching the town?] Hope the rebels will not come in, but if they do Charleston must be destroyed. I have been reading the treach-erous correspondence of the people for the last three months, and I think that our wives and children deserve as much consideration as those of Charleston.
Very respectfully, etc.,
E.P. SCAMMON
Brigadier-General, Commanding

Charleston, May 12, 1863
Colonel HAYES:
I have already given you the "scare." Think there is truth in it. Will probably be over before morning. Meantime, I direct you to have guns in readiness to fire on Charleston. If rebels come in here Charleston shall be destroyed, for it is the work of disloyal citizens.

E. P. SCAMMON
Brigadier-General

HDQRS. FIRST BRIG., THIRD DIV.,
EIGHTH ARMY CORPS,
Camp White, Va., May 12, 1863

Maj. J. P. McILRATH,
Comdg. Twenty-third Ohio Vol. Infantry,
Camp White, Va.,
[AND OTHER SUBORDINATE COMMANDERS]:
SIR: I am directed by General Scammon to have all in readiness for an attack. I do not think it necessary to awaken the men at present, but you will see the sentinels are on the alert, and such preparations are made as will enable you to have your men ready for an attack on the shortest notice. Captain Simmonds and Lieutenant Austin [artillery commanders at Fort Scammon], in case of alarm, will put their guns in position to bear on Charleston.
Respectfully, etc.
R.B. HAYES
Colonel, Twenty-third Ohio Volunteer
Infantry, Commanding

The orders show that General Scammon had in-formation linking Charleston citizens with a rumored raid by Confederate forces of some kind—he does not indicate whether he would be facing regular Con-federate troops or some guerilla band. It is likely that he was preparing for bushwhackers, because the threat of invading Confederate troops had all but disappeared by 1863. Also, it was known that roving bands of guerilla cavalry were operating in the Kanawha, Gauley and Birch river valleys. General Scammon seemed to be alluding to a coordination of bushwhacking opera-tions by people in Charleston, and he intended to destroy the town in retaliation. Fortunately, no at-tack came, and the city was spared.

The bushwhacker war was so intense and feelings so bitter that vindictiveness carried over into the Reconstruction period in West Virginia. Even in the spring of 1865, when it was obvious the war was about over, loyalists throughout West Virginia were holding mass meetings and writing resolutions warn-ing every Rebel, Southern sympathizer, and especially every hated bushwhacker "to return to some other home than that which knew him before, in the days of his innocence."

*From *Fort Scammon and A History Of The Civil War In Charleston And The Kanawha Valley, West Virginia,* Paul D. Marshall & Assoc. Inc., 1986

War in the Kanawha Valley, 1864

Col. Hayes wrote from Camp White in January 1864 that his regiment had reenlisted in the army as a veteran regiment. It was usual, during the wintertime, for enlistment campaigns to be conducted back in the hometowns by members of the regiment. The men took their winter furloughs in company strength. He wrote also about the weather in Charleston: "very very cold indeed"; snow "good sleighing"; the river was frozen over, and they were cut off from navigation. He said: "Provisions were pretty plenty, however, and we felt independent of the weather."

Kanawha Valley action in 1864 began in February in Putnam County, where Confederate Colonel Milton J. Ferguson, colonel of the 16th Virginia Cavalry, tried to disrupt the organization of the county government. (Putnam County, like many other counties, was just beginning to function as an entity of the newly formed state of West Virginia.) The Charleston Federal command quickly responded to the crisis with a small force under Gen. E. P. Scammon. The Union contingent moved all the way to Point Pleasant.

During the return trip (Scammon had insisted on a night journey, against the boat captain's warnings) from Point Pleasant on the steamboat *The Levi*, the boat was boarded, as it stopped to negotiate the treacherous Red House Shoals, by a rebel party of ten Confederate guerillas under the leadership of Maj. James H. Nounnan. General Scammon, four other officers, 25 privates and about $100,000 worth of supplies, were captured. The boat and supplies were burned and the enlisted men were paroled. General Scammon and two aides were sent to Richmond, Virginia, but the general was later paroled and sent to Washington in a prisoner exchange. The Red House Shoals action took place on Feb. 3, 1864.

After a visit home to Ohio in February, Colonel Hayes returned to Camp White with his wife and children on March 11. When in Charleston Hayes' family stayed in a cottage, near Camp White, rented by the Colonel from the prominent Quarrier family. On April 3 he wrote his uncle that he had visited his five posts between Camp White and the mouth of [Big] Sandy and noted that several squads of rebel guerillas had been captured. On April 20 he wrote that he was certain of taking an active part in the summer campaign. On the 24th he wrote to his mother: "We are very busy, and of course happy getting ready for campaigning."

Mrs. Hayes and the boys left Camp White for Ohio at the same time (April 29) Colonel Hayes' command left Charleston in the service of Gen. George Crook, whose mission it was to destroy the salt works at Saltville, Virginia, and to cut and destroy as much as he could of the very important Virginia and Tennessee Railroad. The long, difficult march ended in the Battle of Cloyd's Mountain, near Dublin, Virginia, on May 9, 1864. The battle ended in a resounding Federal victory. The Confederate losses included a mortal wound to their cavalry commander, Gen. Albert Gallatin Jenkins of Cabell County, West Virginia.

The Twenty-third went on from Cloyd's Mountain to battles at Staunton, Brownsburg, Lexington and Lynchburg, Virginia. The army was forced to retreat from Lynchburg before the heavy pressure of reinforcements from Richmond.

They reached Charleston and Camp White on July 1. They rested for ten days and then General Crook's command, including Hayes' Twenty-third, was ordered to the Shenandoah Valley against the army of General Jubal Early. There were sharp contests at Kernstown, Virginia, near Winchester; at Berryville; Opequan; Fisher's Hill (Massanutten Mountain); and the Battle of Cedar Creek, which ended in a great victory for the Union army of General Philip Sheridan.

The victory at Cedar Creek, which ended Confederate occupation of the Shenandoah Valley, wrote the final chapter of a long and bitter war history in the valley. General Grant's grand strategy of 1864 was to lay waste to the "Breadbasket of the Confederacy" on which Lee was so dependent in order to feed his army.

Several initial attempts to oust the rebels met with failure due, for the most part, to an inept Union general. The Battle of New Market on May 15 saw Gen. John C. Breckenridge defeat Gen. Franz Sigel with a patched up force that even included young cadets from VMI. Gen. David Hunter replaced Sigel and was able to penetrate as far as Staunton. But, after a skirmish in June against Jubal Early, Hunter retreated into West Virginia. His retreat saved the wheat harvest for Lee.

After the Battle of New Market, and throughout the subsequent journey through Lynchburg, Staunton, and Lexington, Virginia, and also in his trek through the Kanawha Valley, *General Hunter committed many nonmilitary outrages against the people. Included in the campaign of destruction was the burning of the Virginia Military Institute; the burning of several homes;* and he would have burned the famous health spa hotel at

Gen. David Hunter was born in 1802 and died in 1886. He graduated from the United States Military Academy in 1822; appointed second lieutenant in the Fifth Infantry; promoted first lieutenant in 1828, and became a captain in the First Dragoons in 1833. He resigned his commission in 1836, and engaged in business in Chicago. He re-entered the military service as a paymaster, with the rank of major, in March 1842. On May 14, 1861, he was appointed colonel of the Sixth United States Cavalry, and three days later was commissioned brigadier general of volunteers. He commanded the main column of McDowell's army in the Manassas campaign, and was severely wounded at Bull Run, July 21, 1861. He was made a major general of volunteers Aug. 13, 1861; served under General Fremont in Missouri, and on Nov. 2 succeeded him in the command of the Western Department. In March 1862 General Hunter was transferred to the Department of the South, with headquarters at Port Royal, S.C. In May 1864 he was placed in command of the Department of West Virginia. He defeated a considerable force at Piedmont on June 5. He was brevetted major general, United States Army, March 13, 1865, and mustered out of the volunteer service in January 1866.

White Sulphur Springs had it not been for pleas by Capt. Henry A. duPont of the famous duPont family.

It is ironic that, when young Captain duPont paused at Camp Piatt near present-day Belle, he was standing almost on the spot where, in the future, his family would build the great Belle Works of the duPont Chemical Corporation. General Hunter passed through Charleston in July 1864 on his way to Parkersburg where he entrained his command for the Shenandoah Valley and another campaign against Early. General Hunter was not very well received in the Charleston area by supporters of either side.

Charleston saw no action in 1864 except for the movement of troops through the valley. There was action downstream along the Kanawha in Putnam County. A detachment of Federal troops commanded by Capt. John Reynolds was sent to Winfield in October to occupy the town and provide protection to river transportation. Boats and travelers along the river had been harassed by rebel bushwhackers.

The troops constructed trenches in the town itself and on the hill around the site of the present courthouse. A force of 400 Confederates, under the command of Col. Vincent A. "Clawhammer" Witcher, learned of the Union occupation of Winfield, and determined to attack the town. The attack, carried out at night on Oct. 26, managed to penetrate the center of the defense system but was driven back. During the battle Capt. Phil Thurmond was mortally wounded. The Confederates withdrew to their base on the Mud River, south of Winfield.

The attack on Winfield was not a surprise, except perhaps for the fact it was made at night. Colonel Witcher's activities had been monitored carefully by Union observers from at least October 20, and Captain Reynolds had been notified of his movements. Several dispatches were sent by Col. John H. Oley, Commander at Charleston, to subordinates, on Oct. 22, 1864.

Charleston, W. Va., October 22, 1864

Capt. E. B. BLUNDON
Guyandotte, W. Va.

The rebel Colonel Witcher, with 500 men, was at Marshy Fork of the Coal River yesterday. I think he will try to get down in the Guyandotte country. Fortify and be vigilant. I will watch him and help you all I can if necessary. If he comes fight him at all hazards and act in connection with Winfield and Coalsmouth.

JOHN H. OLEY
Colonel, Commanding Brigade

After the Winfield skirmish, troop movements in the Kanawha Valley ended, except for occasional encounters with bushwhackers, which had become part of ordinary duty. As noted above, Col. John H. Oley, commander of the famous Seventh West Virginia Cavalry, had been placed in command of the Charleston district. He was in charge when news came in April 1865, that General Lee had surrendered.

---

*From *Fort Scammon and A History Of The Civil War In Charleston And The Kanawha Valley, West Virginia*, Paul D. Marshall & Assoc. Inc., 1986

Hunter's Retreat Through the Kanawha Valley

From Staunton an empty wagon train, convoyed by one hundred days men from Ohio, had been started through to the Kanawha. The bushwhackers under our old acquaintance Bill Thurman, had driven them off the road and corralled them in a mountain gorge, where for more than a week they had been in a state of starvation, surrounded by Thurman's men who occupied the ridge and was picking them off as fast as they showed their heads above the rude breastworks constructed. Hearing of their situation a company or two of cavalry went to their relief and compelled Thurman to raise the siege, but did not get close enough to enforce the terms of his violated parole of two years before. The night of this day we encamped at White Sulphur Springs and next morning moved on fording the Greenbrier river past Lewisburg, and by marching all night reached Meadow Bluffs at the eastern base of the Big Sewell mountain about daylight. Here a detail was made on all the companies of our regiment for all the men and horses that could stand a continuous march of 25 miles to report to me, and although we had left the Kanawha 800 strong, all well mounted, we could now muster but twenty-five men with horses fit to attempt this short march and with them I started in advance of the army to escort General Hunter and staff over to the Kanawha, whither they went to hurry out supplies to met the famishing troops.

Arriving at the falls of the Kanawha in the evening Gen. Hunter was in communication by telegraph with the outside world and immediately ordered supplies up from Charleston and sent them on to meet the starving troops, still struggling against the fates in the inhospitable mountains. In a day or two all that was left of men and horses were in and now by easy marches moved down to Charleston where they arrived on the 27th of June. The physical condition of the troops was pitiable in the extreme, but a few good square meals of honest army rations, with a night or two of undisturbed rest and sleep, together with prudent bathing in the limpid waters of the Kanawha, improved and invigorated us wonderfully. I don't know that any official report of the losses of this expedition was ever made, but doubt if 3,000 men and 4,000 horses would cover them. To give an idea of the fatality among the horses, one regiment left 125 on Big Sewell mountain alone, and had but eight left in the entire regiment when they reached Charleston—while others suffered in like proportion.

From *With the Army of West Virginia, 1861–1964, Reminiscences & Letters of Lt. James Abraham, Pennsylvania Dragoons, Company A First Regiment, Virginia Cavalry,* Compiled by Evelyn Abraham Benson, Lancaster, Pa., 1974.

"You will please have Robert Brown of C. G. 11th Regiment, Va. Vol. Infantry arrested & sent to these Head Quarters. He has been guilty of committing outrages by destroying the fences of Phillip Epling who can give you information so that you can have him caught. R. Yount Lt. Col. Commg. Post"

Head Quarters Post Charleston West Va May 1864

Commanding Officer Coles Mouth Va

You will please have Robert Brown of C. G. 11th Regt Va Vol Infy arrested & sent to these Head Quarters. He has been guilty of committing outrages by destroying the fences of Phillip Epling who can give You information so that You can have him caught. R. Yount Lt Col Commg Post

The 23rd Ohio Volunteer Infantry band in Charleston in 1863. It is one of the few photos taken during the war in Charleston known to exist. The building in back of the band is the Bank of Virginia, at the corner of present Virginia and Summers streets. It was burned during Lightburn's retreat in 1862. RBHL.

A stock certificate of the Great Kanawha Valley Oil Company, June 1865. No other information is available on this venture.

CHARLESTON
Daily Bulletin.

Vol. I. }
No. 11. }

Charleston, W. Va., June 13, 1864.

{7 o'clock,
{Evening.

MOORE & BROTHER, Publishers.

OFFICE, "BANK OF THE WEST" BUILDING.

Terms (in advance).

For Single copy, ...   5 cents.
For One Week, delivered by carrier, .   . 25 "
For One Month, "   "   . $1 00

RATES OF ADVERTISING.

One Square (10 lines), or less, one insertion, 50 cents.
  "   "   "   one week, $1 00
  "   "   "   one month,   3 00
Two Squares, one month, ... 5 00

Monday, June 11, 1864.

Thurman's Guerrillas.—Their Captain Captured.

The train, consisting of half a dozen old and empty wagons, sent out with 35 men of the O. N. G. as a guard, for the purpose of gathering up the broken-down wagons on the road between Gauley and Meadow Bluff, met with misfortune on Friday morning near Sewall Mountain.

The wagon-master, Dan. Lee, of the 12th Ohio, hearing of the presence of Thurman's guerrillas when at the foot of Sewall, proposed to leave the train with a guard, and with the rest of the force go up to the top and see if any were to be found. Some eight of them, having horses, were in the advance, —the infantry following as fast as they could, when, arriving on top of Big Sewall, they saw two mounted rebels and charged upon them. But these were a bait, and four of our boys, including the wagon-master, soon found themselves surrounded by Thurman's entire company, who, upon their refusing to surrender, fired upon them. Their horses being shot under them, three of them escaped to the woods and got away safely; but nothing has been heard of the fourth, Private Joseph Johns, of Comp. H, 12th O. V. I. We fear he was killed.

The guerrillas followed as far as where the train had been left, but being disappointed in finding any plunder there, they burned the wagons and returned.

Two of the members of the 23d O. V. I., belonging to the train, having secreted themselves until the following day, were fortunate enough, having merely side-arms, to capture Capt. Bill Thurman, the notorious leader of this gang. He is now here under guard.

☞ Late intelligence from General CROOK is to the effect that he is operating disastrously to the rebels. How and where, it is not deemed proper to publish at present.

BY TELEGRAPH.

THIS MORNING'S DISPATCHES.

Grant changing his Base.

Official Advices from Sherman.

FURTHER FROM BUTLER.

Gen. Smith driving Marmaduke.

A Battle near Cynthiana, Ky.

CINCINNATI, June 13.

News from Grant's army indicates that the base of supplies is being transferred from White House, and by this time probably the Headquarters are on James river.

This means an attack on Richmond from the south-west. The position, if secured, will enable Grant to cut off supplies and thus compel Lee to come out and fight.

The movement across the James river may bring on a battle, but there had been no engagement up to Saturday afternoon.

Reports from Gen. Hunter show, as expected, his victory over Jones to have been much more important and complete than reported in the rebel papers. Crook and Averill formed a junction with him at Staunton.

Our correspondent with Gen. Sherman's

army furnishes a corrected account of the movements up to June 5th, inclusive.

Official advices from Sherman are up to yesterday morning. The two armies were then confrontong each other.

Gold fell in New York on Saturday from 98 to 93½, and closed dull at 94.

Affairs in Gen. Butler's department are progressing favorably. On Thursday Gen. Gilmore advanced on Petersburg, but withdrew when hearing that the rebels had been forewarned of his approach. Kautz's division, however, dashed forward and forced the outer works, capturing a few prisoners and a gun or two. Not being supported, he was obliged to fall back, which he did with trifling loss, taking his prisoners and guns with him.

While this was going on, the gunboats on the Appamatox vigorously shelled the town, and a detachment of men tore up three or four miles of the Petersburg and Richmond railroad.

Gen. W. S. Smith has cleared away the obstructions to the navigation of the Mississippi at Columbia, Ark.,—a battery, supported by the rebel Marmaduke with several thousand men. Gen. Smith landed his forces the 5th inst., silenced the rebel artillery, and drove the enemy away. They succeeded in getting beyond a lake which prevented speedy pursuit. Gen. Smith's loss was 20 killed and 70 wounded. The rebels lost about the same number.

Gen. Hobson with 600 men fought Morgan's whole force near Cynthiana for seven hours, but was finally compelled to surrender or be demolished.

Charleston Daily Bulletin.

Charleston, W. Va., June 13, 1864.

☞ CARELESS SHOOTING.—We are pained to learn that some members of the Cavalry encamped on the other side of the river have been very careless, or else reckless, in discharging their fire-arms. Mr. Gibson, who lives on that side, while cutting some wood yesterday, barely escaped losing his life from one of their bullets.

☞ THE STEAMER VICTRESS.—We are glad to see this fine boat at the wharf once more. She has just been thoroughly repaired and neatly re-painted, and looks as bright as a new postal currency! If you don't believe it, go and see for yourself.

☞ The New York gamblers finally succeeded in raising gold to 100 prem., but it is falling faster than it went up. We hope they are satisfied.

General Hunter's Guerrilla Circular.

The guerrillas have been very bold and destructive in their murderous attacks of late. Besides those instances already given several others have come under our notice. Mr. Wm. Nixon, ex-sheriff and enrolling officer of Wayne county, was taken out of Ceredo and coolly shot down, but a few days ago, by the renegade land-pirate Jim Smith.

We give below Gen. Hunter's guerrilla Circular, which, we are authorized to say *will be carried out to the letter* by the Commandant of the valley. Let our rebel friends (?) keep their eyes open if they do not wish to lose their property, or personal liberty, *or both!*

HEAD QUARTERS DEPARTMENT W. VA.,
In the Field, Valley of Shenandoah,
MAY 24, 1864.

Circular:

Your name has been reported to me, with evidence that you are one of the leading Secession Sympathizers in this Valley, and that you countenance and abet the Bushwhackers and Guerrillas who infest the woods and mountains of this region, swooping out on the roads to plunder and outrage loyal residents, falling upon and firing into defenceless wagon-trains, and assassinating soldiers of this command who may chance to be placed in exposed positions. These practices are not recognized by the laws of war of any civilized nation, nor are the persons engaged therein entitled to any other treatment than that due, by the universal code of justice, to pirates, murderers and other outlaws.

But from the difficulties of the country, the secret aid and information given to these bushwhackers by persons of poor class, and the more important occupation of the troops under my command, it is impossible to chase, arrest and punish these common marauders as they deserve. Without the countenance and help given to them by the rebel residents of the Valley, they could not support themselves for a week. You are spies upon our movements, abusing the clemency which has protected your persons and property, while loyal citizens of the United States residing within rebel lines, are invariably plundered of all they may possess, imprisoned, and in some cases put to death. It is from you and your families and neighbors that these bandits receive food, clothing, ammunition and information; and is from their secret hiding places in your houses, barns, and woods, that they issue on their missions of pillage and murder.

You are, therefore, hereby notified that for every train fired upon, or soldier of the Union wounded or assassinated by bushwhackers in any neighborhood within the reach of my cavalry, the house and other property of every secession sympathizer residing within a circuit of five miles from the place of the outrage, shall be destroyed by fire; and that for all public property jayhawked or destroyed by these marauders, an assessment of five times the value of such property will be made upon the secession sympathizers residing within a circuit of ten miles around the point at which the offense was committed. The payment of this assessment will be enforced by the troops of the department, who will seize and hold in close military custody the persons assessed until such payment shall have been made. This provision will also be applied to make good, from the Secessionists in every neighborhood, five times the amount of any loss suffered by loyal citizens of the United States from the action of the bushwhackers whom you encourage.

If you desire to avoid the consequences herein set forth, you will notify your guerrilla and bushwhacking friend to withdraw from that portion of the Valley within my lines, and to join—if they desire to fight for the rebellion—the regular forces of the Secession army in front or elsewhere. You will have none but yourselves to blame for the consequences that will certainly ensue if these evils are permitted to continue.

This circular is not sent to you for the reason that you have been singled out as peculiarly obnoxious, but because you are believed to furnish the readiest means of communication with the prominent Secession sympathizers of your neighborhood. It will be for their benefit that you communicate to them the tenor of this circular.

D. HUNTER, Maj.-Gen. com'dg.

J. D. MITCHELL, Lieut. & A. Post Adj't.

BY TELEGRAPH.

Fort Darling Captured!

The Enemy Repulsed by McPherson, leaving 2500 on the field!

Morgan Repulsed at Frankfort!

BURBRIDGE ROUTS MORGAN AT CYNTHIANA!

From General Butler.

NEW YORK, June 13, 1864.

The Herald's correspondent sends the following:

WHITE HOUSE, June 11.—News from the front this morning is most cheering. Two officers who have just arrived here, bring the joyful news of the capture of Fort Darling.

An order conveying this intelligence was read to the army last evening, and the cheers of our soldiers could be heard miles around.

From General Sherman.

NEW YORK, June 13, 1864.

The attack on McPherson proved very disastrous to the enemy. The rebels came on in two divisions, with great resolution, but were met with a very destructive fire of artillery and musketry. The fight continued for nearly an hour, when the enemy retreated, leaving the field covered with their dead and wounded, numbering nearly 2500.

After five days' fighting, principally on his own hook, McPherson has closed upon our right wing, enabling us to make important movements.

Latest from Morgan.

LOUISVILLE, KY., June 12.

Dr. Whaler, U. S. Mail agent, who has been at Frankfort during the seige, left there at 4:30 this morning, and reports that fighting commenced at 6 o'clock Friday evening, lasting till dark, and continued at intervals during the night; the enemy approaching from Georgetown, in two forces, aggregating 1200. Seven hundred entered Old Frankfort, and five hundred entered New Frankfort. They had no artillery.

A small 4-pounder had been placed below the fort to protect our rifle-pits, which was captured by the rebels, but was subsequently retaken.

On Saturday firing continued from 7 o'clock in the morning to 3 o'clock in the afternoon, with short intervals. The rebels made two demands during the day for the surrender of the fort, both of which were refused by Col. Monroe of the 22d Kentucky, commanding the fort; the rebels abandoned the attack at 4 o'clock on Saturday afternoon.

By 7 o'clock in the evening they were moving eastward. Our loss was 6 wounded, one of whom seriously. The rebel loss is unknown.

The fort was garrisoned by 150 Federals, only 12 of whom were soldiers.

No injury was done to Frankfort, except burning the barracks and a bridge 3 miles northward.

Capt. Dickson, of Gen. Burbridge's staff, telegraphs to Gen. Ewing at Lexington at 9:35 P. M., that Burbridge completely routed Morgan's command at Cynthiana this afternoon.

Photos and drawings from the Civil War era in Charleston are rare. This rather idyllic scene was painted in 1863 by Margaretta Doddridge, from a vantage point near what is now Morris Street. It shows a Union army camp (Camp White) on the south side of the Kanawha River, around the Ferry Branch outlet just downriver from the present C&O railroad station. PHC

Officers were fortunate to be joined by their wives in camp. This unidentified Federal group with its elaborate log hut is typical of the Kanawha Valley campaign. PHC

After the War

Grant's official dispatch read: "General Lee surrendered the Army of Northern Virginia this afternoon [April 9, 1865—Palm Sunday] on terms proposed by myself."

A formal celebration was held in Charleston under the direction of Capt. Elias Powell. There was a parade, led by Colonel Oley, which featured an old hearse with the word "Secession" painted on its sides, and a wagon with Jefferson Davis hanged in effigy from a "sour apple tree." The parade moved to the top of Cox's Hill at the north end of present Capitol Street, where guns from Battery A, First West Virginia Artillery, fired several volleys to celebrate the fall of the Confederate forces. A prayer was read by Chaplain E. W. Gregg of the Seventh West Virginia Cavalry, followed by an address by Col. James H. Ferguson.

Colonel Oley disbanded his Seventh West Virginia Cavalry and other Federal units, and his office handled the parole of almost 5,000 Confederate soldiers.

The Twenty-third Ohio Volunteer Infantry was a remarkable military unit not only with respect to the quality of leadership during its history as a fighting regiment, but also after the war ended. Its first commander, William S. Rosecrans, was not only a successful engineer and businessman before the war, but served as a minister to Mexico, congressman from California, and Register of the Treasury after the war. Eliakim P. Scammon later represented the United States as consul at Prince Edward Island. James Comly, who followed Hayes as commander, became the American minister to Hawaii. Stanley Mathews served a term in the U.S. Senate and became a justice of the Supreme Court in 1881. William McKinley entered the service as a private in 1861 and rose to the rank of brevet major. Rutherford B. Hayes, in his early service, said he liked the title "Colonel" best of all, but his valor on the field of battle earned for him the rank of brigadier general and he finally came out of the war experience as Major General R. B. Hayes. At the end of his life he responded to the title "General" far more warmly than he did to "Mr. President."

---

*From *Fort Scammon and A History Of The Civil War In Charleston And The Kanawha Valley, West Virginia*, Paul D. Marshall & Assoc. Inc., 1986

Paroled prisoner's pass of Henry M. McCown of Co. A, 36th Inf. CSA. H. M. McCown was born near Buffalo in 1838. He died in 1913 at Ashland, Ky. He was a grandson of Malcolm C. McCown of Captain Arbuckle's Company at Point Pleasant, 1774. Malcolm later became a pioneer settler in the vicinity of Buffalo. He was the son of Frances McCown, who migrated from Ireland to August County, Va., in 1740. wvsa

A parole document issued at Charleston on June 12, 1865, to an ex-Confederate soldier. wvsa

Echoes of Glory
Legacy of the War

"Soldiers Always"
Veterans Organizations

After the Civil War veterans of both armies organized groups similar to today's American Legion. In 1866 the main Union organization, the Grand Army of the Republic, was formed. In 1889 the United Confederate Veterans was formed.

The original UCV camp in Charleston was the Stonewall Jackson Camp, No. 878. Ex-Kanawha Riflemen member Thomas L. Broun was named commander in 1908.

Thomas Broun's son Fontaine formed the Camp Thomas L. Broun, United Sons of Confederate Veterans, in Charleston in 1900. (The Sons of Confederate Veterans was organized at Richmond, Va., in 1896.) There is still a national Sons of Confederate Veterans organization with the closest camp in Guyandotte. The organization was named for Big. Gen. Robert S. Garnett, the first general officer to be killed in the war on July 13, 1861, at Corrick's Ford near Parsons in Tucker County.

The Kanawha Valley's last Confederate veteran was Eli J. Carte of 313 Bibby Street who was 100 years old in 1940. He died in 1941.

The Grand Army of the Republic, of course very strong in the north and west, was also strong in the northern cities of West Virginia. The headquarters for the Department of West Virginia, G.A.R. was established in 1882. In 1892 the G.A.R. had several posts in the Kanawha Valley. Post 73 in Charleston had 100 members, Post 90 in St. Albans had 22 members, Post 95 in Clendenin had 30 members, Post 3 in Coalburg had 45 members, Post 11 in Point

Pleasant had 26 members, Post 102 in Hurricane had 13 members and Post 114 in Letart had 23 members.

In the 1940s the national president of the Ladies of The Grand Army of The Republic, Gladys W. Newton, was from Charleston. Today the Sons of Union Veterans is a national organization.

The United Daughters of the Confederacy was formed in 1894. The Charleston chapter, still in existence, was formed soon afterwards. In 1900 the chapter had 103 members.

UNITED DAUGHTERS
of the
CONFEDERACY, (Inc.)

Thomas L. Broun in later life.

The Sons of Confederate veterans named their camp after Thomas L. Broun, prominent Charleston attorney and member of the Kanawha Riflemen.

Commandant,
W. D. PAYNE.

Adjutant,
C. C. LEWIS, Jr.

Treasurer,
C. G. PEYTON.

Historian,
W. E. R. BYRNE.

1st Lieut. Com.,
JOHN BAKER WHITE.

2d Lieut. Com.,
JOEL H. RUFFNER.

Surgeon,
Dr. V. T. CHURCHMAN.

Quartermaster,
RANDOLPH T. CARMICHAEL.

Color Sergeant,
T. R. TEMPLIN.

Chaplain,
Rev. H. G. WILLIAMS.

HEADQUARTERS
CAMP THOMAS L. BROUN,

United Sons of  Confederate Veterans,

CHARLESTON, WEST VIRGINIA.

Charleston, W. Va. Nov. 18, 1902 190

Mr Nevell S. Bullett,

Commander Kentucky Division U.S.C.V.,

Louisville, Ky.

My Dear Bullett:-

Pardon my not acknowledging receipt of yours of the 17th inst, before this, pressure of work prevented.

I have not yet received the pins U. S. C. V. which you said would be received in about two weeks from October 17th. Please hurry same up.

I am sorry you could not come and make Angus and myself a visit this fall, but shall hope to have that pleasure some future day. I certainly hope that both Angus and I can join you boys and go with you to the Reunion next year. At what time will the Reunion be held?

Yours very truly,

Fontaine Broun

INCORPORATED UNDER THE LAWS OF
WEST VIRGINIA

Stonewall   **Jackson**

FULL-PAID   NON-ASSESSABLE

This Certifies That   CAMP   is the owner of

John Clark

One Share of the Capital Stock of STONEWALL JACKSON CAMP, not subject to Assignment, Sale, Pledge or Transfer.

In Witness Whereof, the duly authorized officers of this Corporation have hereunto subscribed their names and caused the corporate seal to be hereunto affixed, at Charleston, W. Va., this 11 day of May 1908.

J. W. Vickers, Secretary   A. T. Wilcox, President

APPLICATION FOR MEMBERSHIP.

To the COMMANDER, OFFICERS, COMRADES, CONFEDERATE VETERANS
OF

STONEWALL JACKSON CAMP No. 878, U. C. V.
Charleston, West Virginia.

Desiring to be of some service to the charitable object as set forth in your By-Laws, I offer myself for membership, and if elected will comply with the By-Laws of the same. Herewith please find enclosed $1.00 for Muster Fee, which I will expect returned if I am rejected.

I was born 7th day March 1843 in Rockbridge County.

State of ____ Enlisted 5 day November 1862

Where at Winchester Va By whom Capt Alex G McChesney

Company 7 Regiment 1st B=11th Va Cav Brigade Robinson
afterward 11th Va Cav Wm E Jones (Laurel)
was of V M I Jany 62. Will Cadet as McDowell Battle

Rank Private Promoted

Transferred August 9 1863 to 11th Cavalry Va Captured

Remarks to Co C 14th Va Cavalry Jenkins afterward McCausland Brigade ...

Honorably Discharged, Paroled was on sick leave at close of war, parole at Staunton sometime in June.

I hereby affirm that the above is a true record, as witness my hand and signature.

Signed (Full name signed) James L McChesney

Vouched for by member of this camp.   P. O. Charleston W V

Date May 5th 1907

Refer to war papers &c
Genl Jno McCausland Grimms Landing W Va
H H Stephenson Montery Va

Veterans Groups in the
Kanawha Valley as of 1911

Blunden Post, No. 73, G.A.R. Meets every 2nd and 4th Saturday of each month in I.O.O.F. Hall

Blunden Women's Relief Corps, No. 6, Meets every other Tuesday in I.O.O.F. Hall.

George Crook Post, No. 3, G.A.R. Meets 1st and last Friday of each month in I.O.O.F. Hall.

George Crook W.R.C., No. 16. Meets 1st and 3rd Friday in I.O.O.F. Hall.

Camp Thomas L. Broun, United Sons of Confederate Veterans, No. 193. Meets at the call of the commandant at the office of Payne & Payne.

Camp R. E. Lee 887, United Confederate Veterans. Meets on first Saturday in I.O.O.F. Hall.

Charleston Chapter ISI United Daughters of the Confederacy. Meets 2nd Monday of each month at members' residences.

Stonewall Jackson Camp 878, United Confederate Veterans. Meets 2nd Monday of each month at I.O.O.F. Hall.

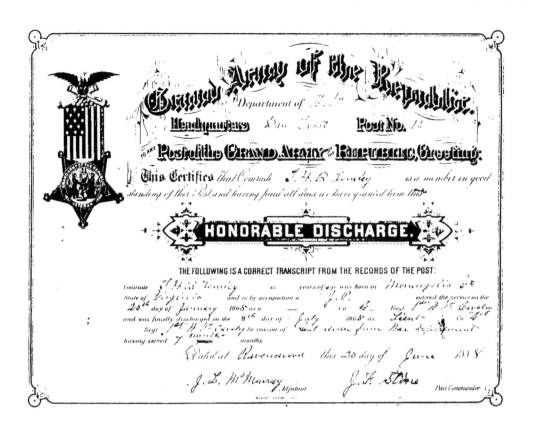

DEPARTMENT OF WEST VIRGINIA,

GRAND ARMY OF THE REPUBLIC.

I. M. ADAMS,
DEPARTMENT COMMANDER.

A. J. CHARTER,
ASSISTANT ADJUTANT GENERAL.

I. W. WATSON,
MORGANTOWN.
JUNIOR VICE COMMANDER.
T. H. MARKS,
WELLSBURG.
DEPARTMENT CHAPLAIN.
TAYLOR RICHMOND,
R. F. D. 3, FAIRMONT.
MEDICAL DIRECTOR.
J. H. BROWNFIELD,
FAIRMONT.
QUARTERMASTER GENERAL.
J. S. MORGAN,
RAVENSWOOD.
JUDGE ADVOCATE.
COL. HENRY HAYMOND,
CLARKSBURG.

Ravenswood, W. Va., July 30, 1907.

HONORABLE W. M. O. DAWSON,

    GOVERNOR OF WEST VIRGINIA,

        CHARLESTON, WEST VIRGINIA.

    SIR:—

        I have been instructed by the Department of West Virginia assembled at Fairmont, W. Va. last May to request you to deliver to competent persons whom we will select (with your approval) a few of the old but valued flags under which we fought from 1861 to 1865 so that they may be taken to Saratoga, N. Y. to the National Encampment, which meets there September, 9th, 1907.   We will care for them and return them as good as when we received them, they are ours we bought them with the lives of many of our comrades and not any one would care for them as we shall.   Hoping for a favorable answer, I am,

        Yours truly,

        —————————————————————
        Department Commander

CAMP THOS. L. BROUN

Organized Here Last Night by Sons Confederate Veterans.

FONTAINE BROUN ELECTED COMMANDANT OF THE CAMP.

Constitution and By-Laws Adopted and Applications For Membership Received --Another Meeting--Tuesday.

Camp Thomas L. Broun, United Sons of Confederate Veterans, was organized here last night. Fontaine Broun was elected commandant. The election of the other officers was deferred until another meeting, to be held Tuesday night.

The meeting last night was held at the law office of Broun & Broun. W. Dallas Payne was elected temporary chairman and R. P. Flournoy temporary secretary. Several old veterans from Stonewall Jackson camp, were present and addressed the meeting.

Angus W. McDonald, of Louisville, Ky., who had been detailed here by Commander-in-Chief Briscoe Hindman, of Louisville, Ky., for that purpose, assisted in the formation of the camp.

A constitution and bylaws were adopted, and applications for membership were received and filed. It is expected that the camp will be composed of from 75 to 100 members.

After appointing various committees the meeting adjourned.

The meeting Tuesday night will be held at 8 o'clock at the law office of Payne & Payne, in the Kanawha valley bank building.

The formation of this camp, the first of the kind organized in this city, is the outcome of a movement started several weeks ago, at the suggestion of several local veterans, who desired to have their own camp in a way confirmed. The objects of the Sons of Confederates are similar to those of the senior organization. They are stated in this way in the preamble to the constitution:

"The objects and purposes of this organization shall be strictly 'historical and benevolent.' It will strive:

"To unite in one general confederation all associations of Sons of Confederate Veterans, soldiers and sailors now in existence, or hereafter to be formed, and to aid and assist the United Confederate Veterans and all veteran camps.

"To cultivate the ties of friendship that should exist among those whose ancestors have shared common dangers, sufferings and privations.

"To encourage the writing by participators therein of accounts, narratives, memoirs, histories of battles, episodes and occurrences of the war between the States.

"To gather authentic data, statistics, documents, reports, plans, maps and other material for an impartial history of the Confederate side; to collect and preserve relics and mementoes of the war; to make and perpetuate a record of the services of every member of the United Confederate Veterans, and all living Confederate veterans, and, as far as possible, of those of their comrades who have preceded them into eternity.

"To see that the disabled are cared for; that a helping hand is extended to the needy, and that needy Confederate veterans' widows and orphans are protected and assisted.

"To urge and aid the erection of enduring monuments to our great leaders and heroic soldiers, sailors and people and to mark with suitable head stones the graves of Confederate dead wherever found.

"To instill into our descendants a proper veneration for the spirit and glory of our fathers, and to bring them into association with our confederation, that they may aid us in accomplishing our objects and purposes, and finally succeed us and take up our work where we may leave it."

Charleston Gazette, June 4, 1914.

CONFEDERATES PAY TRIBUTE TO DEAD COMRADES

Members of Lee and Jackson Camps Decorate Graves In Cemetery

HONOR JEFFERSON DAVIS

Stonewall Jackson Statue Is Draped With Flowers By Veterans

The Charleston Confederate veterans, assisted by their wives, children and friends, yesterday paid tribute to their dead comrades and honor to the memory of Jefferson Davis, whose birthday they observed.

Shortly after 10 o'clock nearly all of the 110 members of Camp R. E. Lee, No. 357, Confederate Veterans and "Stonewall" Jackson camp, Confederate Veterans, composed of 107 members, held brief but impressive exercises. The Lee camp met in the Odd Fellows' hall, while the Jackson church. A large delegation of relatives and friends of the veterans were present at both places.

The principal address made to the Lee camp was by Attorney C. L. Smith, who spoke on Jefferson Davis, paying the Confederate president a glowing tribute as a man and executive officer during the stormy period of the great Civil war.

At the conclusion of the exercises the wives and daughters of the veterans served ice cream and cake to all present. During the past year the Lee camp has lost seven members by death.

The "Stonewall" Jackson Camp held their exercises on the hill near Spring Hill cemetery, where an address was made by the Rev. Dr. Chambliss, of Newark, N. J., who was a chaplain in the Confederate army. In the course of his talk he said that on a number of occasions during the war he had General Robert E. Lee among his listeners.

Ceremonies were held by the Jackson camp in the state house yard with Col. S. S. Greene presiding. He delivered a brief address, followed by introducing the Rev. Dr. G. W. Banks of the Southern Methodist church, who delivered the memorial address.

Statue Is Decorated

An impressive and pretty feature of the exercises was the decoration of the statue of "Stonewall" Jackson. The Daughters of the Charleston Chapter presented a number of veterans with the Cross of Honor for bravery and honor as soldiers.

Following the decoration of the graves, the veterans and their relatives and friends returned to the Baptist church, where luncheon was served.

There have been about five deaths in the Jackson camp during the past five years, making a total of more than a dozen members of the two local camps to join the army of the majority in twelve months.

All the men who wore the gray to die in this community are buried in Spring Hill cemetery, where veterans and relatives from both camps decorated their graves in the morning with flowers of remembrance and buds of honor.

Counting the Rebs

A good historian's goal is the presentation of the truth based on facts. Without access to the facts writers sometimes resort to mere conjecture or wishful thinking that can border of propaganda. Author Jack L. Dickinson, in his 1995 book *Tattered Uniforms and Bright Bayonets*, sheds new light on the actual numbers of men from western Virginia who fought for the Confederacy. After an exhaustive ten-year study, Dickinson has determined that early historians drastically *underestimated* the total.

With a careful methodology, Dickinson states that more than 17,000 served the Southern cause, which is in sharp contrast to much lower figures given by Boyd Stutler and others. The 17,000-plus total lays to rest the long-held myth of overwhelming Union sentiment in the area of Virginia that was to become the state of West Virginia.

Other scholars, including James Linger in 1989, agree with Jack Dickinson and until someone can prove otherwise the 17,000-plus total is very convincing. The Civil War itself was a difference of opinion and now—as then—the accuracy of totals on each side is debatable.

THE CHARLESTON GAZETTE.

THURSDAY, JUNE 4, 1908.

Graves of Soldiers of the South are Flower Decked.

Centenial of Davis' Birth and Memorial Day Services Held.

The one hundredth anniversary of the birth of Jefferson Davis and the Annual Confederate Memorial Day were yesterday celebrated impressively by the United Confederate Veterans, the Daughters of the Confederacy and the Sons of Veterans.

The Veterans from Charleston were joined in the ceremonies by old soldiers from Malden, St. Albans and other points in the country, while Putnam county also contributed a large number of her sons who had worn the gray.

Many of the visiting Veterans brought their families with them, and these, with the large number of Sons and Daughters who participated, made the procession, which formed at the corner of State and Summers streets, and moved promptly at 10 o'clock, the largest marching body that has done honor to the dead but immortalized chivalry of the South.

The procession, after the formation, marched to the grave yard and there with Mr. S. S. Green presiding, impressive memorial services were held under the auspices of the Stonewall Jackson Camp, United Confederate Veterans.

Dr. Ernest Thompson, with an earnest and eloquent prayer, opened the services.

This photo was taken on the steps of the old capitol building in downtown Charleston about 1910 by the well-known Gravely Photographers. The men are members of R. E. Lee Camp No. 878, UCV of Charleston. Names are as follows: (1) James L. Kelley, Co. D, 8th Va. Cavalry; (2) D. C. Lovett, Co. C, 8th Va. Cavalry; (3) John Henry Wilson, Co. H, Kanawha Riflemen, 22nd Va. Infantry; (4) Preston Martin, Co. C, 36th Battalion, Va. Cavalry; (5) Veto Farrar, Co. A, 36th Va. Infantry; (6) George S. Chilton, Co. E, 22nd Va. Infantry; (7) J. Press Lanham, Co. A, 22nd Va. Infantry; (8) William J. Thomas, Jackson's Va. Battery; (9) James Z. McChesney, Co. F, 11th Va. Cavalry, transferred to Co. C, 14th Va. Cavalry; (10) James L. Jones, Co. A, Huger's Battery of Artillery; (11) John F. Ballard, Co. E, 22nd Va. Infantry; (12) William C. Hopkins, Co. E, 36th Battalion, Va. Cavalry; (13) Henry M. Brown, Co. B, 44th Va. Infantry; (14) John N. Hutchinson, Co. C, 36th Va. Infantry; (15) N. O. Sowers, Co. I, 2nd Va. Infantry, Stonewall Brigade; (16) Albert J. Wallen, Co. D, 12th Georgia Battery Artillery; (17) Elisha H. Merricks, Lowery's Va. Battery; (18) George W. Mays, Co. K, 24th Va. Infantry; (19) Pleasant Bailey, Co. A, 22nd Va. Infantry; (20) Samuel Motley, Co. A, 8th Va. Cavalry; (21) Henry D. McFarland, Kanawha Riflemen, Co. H, 22nd Va. Infantry. The man standing at the far left is Dr. Will Tompkins, the boy is Eugene Gleason. WVSA

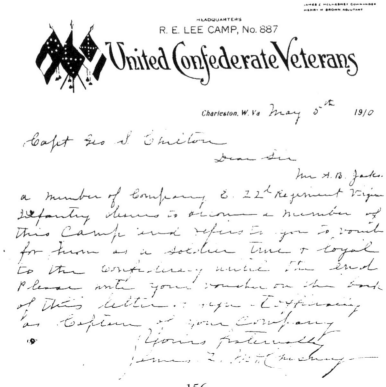

- 156 -

Reunion of the 23rd Regimental band and the Ohio Volunteer Infantry Regiment in Lakeside, Ohio, in August 1885. This regiment spent considerable time in the Kanawha Valley in the latter part of the war. RBHL.

Members of the R. E. Lee Camp, No. 878 of the United
Confederate Veterans, parade down Capitol Street on Sept.
27, 1910, on their way to unveil the Stonewall Jackson
monument at the state capitol lawn. The arcade building is
in the background, the old post office is to the extreme
right. James Z. McChesney, commander of the camp, is on
the right with the gray coat. He was from Rockbridge
County, Va., and graduated from Washington College in
1861. That same year he enlisted in the Confederate army.
McChesney retired from service with a saber wound in
March 1865. He and his wife, Lucy, moved to Charleston in
1871 and engaged in business and selling life insurance until
his death in 1922 at age 79. He held prominent positions in
both the local and national United Confederate Veterans.
WVSA, JAN HUTCHINSON ABBOTT COLL.

Col. Henry C. Dickinson, a leading citizen of Charleston,
was born in 1830 and died in 1871. He graduated from
Hampden-Sidney College with a law degree and practiced
in Bedford County, Va. He joined the Confederate army
with the 2nd Virginia Cavalry. Dickinson was captured
and imprisoned with 600 others (known as the
"Immortal 600") for a time in the harbor of Charleston,
S.C. Because as an ex-Confederate he could not practice
law, he joined his father in the Kanawha Valley salt
business. Dickinson was a founder of the Kanawha Valley
Bank and was mayor of Charleston when he died. His
wife, Sally J. Lewis, was the daughter of John D. and Ann
Lewis. John was the son of Col. Charles Lewis, a hero of
the Battle of Point Pleasant in 1774.

H. C. Dickinson's monument in the Old Circle, Spring Hill Cemetery in Charleston.

Civil War Veterans of Putnam County

West Virginia was born a border state during the Civil War and Putnam County was already a border county. This meant that the state, the county, and even families were widely separated in personal convictions.

For instance, George W. Smith of Eighteen-Mile Creek enlisted in the Union Army while two of his brothers enlisted in the Southern Army. Although he returned and lived in the same neighborhood after the war, he was completely ostracized by his family except for his mother.

Dr. Thomas C. Atkeson who was raised near Buffalo wrote that there were six brothers in his father's family and they were divided evenly; three for the North and three for the South. Not only did they differ on the issues of the war, they were equally divided in politics. In fact, the Union side of the clan even changed the original spelling of the family name from "Atkeson" to "Atkinson."

One flowery writer recorded the following far-fetched postlude to the Civil War, referring to Putnam and Mason Counties. "They are farmers today, statesmen tomorrow, and soldiers always. The performance of the Valley men lent honor to their ancestral heritage and maintained the soldiery of the Great Kanawha Valley. With the return of peace these men came home, laid by their military trappings, donned the citizen's garb and united in an effort to secure the intellectual and industrial development of their beautiful valley."

Possibly some of this epilogue was true except the "live happily ever-after" part. Actually, the returning Confederate veterans not only faced the resentments of many of their neighbors, they were confronted with a demoralizing and devastating state law passed on May 24, 1866, which read as follows:

> No person who, since the first day of June, 1861, has given or shall give voluntary aid or assistance to the rebellion against the United States, shall be a citizen of this state, or be allowed to vote at any election held therein, unless he has volunteered into the Military or Naval service of the United States, and has been or shall be honorably discharged therefrom.

The ratification of this law effectively disenfranchised between ten and twenty thousand West Virginia citizens and denied them the right to practice law, sit on juries, sue in courts, and teach in public schools. This highly restrictive decree remained in effect until May 26, 1871, when it was repealed by the controversial Flick Amendment.

General Edward Ord and family. Mary Thompson was a daughter of Robert Thompson, an attorney in Charleston. When she was 14 her mother died and she came to Coalsmouth to live with her grandmother Mrs. Philip Thompson. Her father later married Elizabeth P. Early of Buffalo, a sister of Gen. Jubal Early. In 1849 Robert Thompson and his family moved to San Francisco, Calif. Here Mary met Maj. Gen. Edward C. Ord of the U.S. Army and they were married. Mary Ord later became the target of Mary Todd Lincoln's public outbursts of jealous rage.

Unreconstructed Rebel BY SHIRLEY DONNELLY

Vignettes of General McCausland

1. He never talked of the War Between the States until late in life. He rarely attended the re-unions of the Blue and Gray. The General always gave as his reason for not attending them that he did not want to give them a chance "to crow over me."

2. General McCausland voted Democratic but said that "when the Democrats get in they always act a fool."

3. When President William McKinley was assassinated at the Pan-American Exposition at Buffalo, New York, by the anarchist, Leon F. Czolgosz of Kanawha City, West Virginia, on September 9, 1901, McCausland said, "I'm glad of it! He was one of General Hunter's staff."

4. The General liked flowers but the rose was the only flower of which he knew the name. Among the trees, the elm was the favorite of the patriarch of Pliny.

5. When it came to temper, the man who burned Chambersburg was quick. It was soon over, though. As a boy he was in a fight almost daily. When twitted about his temper, the General would always say that, "A lively colt makes a high-stepping carriage horse." He liked dogs and horses and had a way with horses that characterized him as a cavalry leader. McCausland also liked turkeys and disdained chickens. Said he, "A chicken is a common thing, but the turkey is an aristocratic bird!"

6. General McCausland was kindly mannered with his children. Knowing history as he did, the General remarked that, "The Chinese govern their children with love and I like that."

7. A blue-gray eye was the eye of General McCausland. His was an affectionate disposition among those who were close to him. He was always quick with an answer and a ready wit. Getting off a train one day, the Conductor asked, "Can you get down those steps, General?" Replied the old Confederate leader, "Yes! I can climb the golden stairs!" His mind was clear right down through the last year of his life. He had the visage of an eagle. His eyes changed to be almost black when he was angered.

8. All during the four long years of the 1861–65 strife, McCausland never came home a single time. "Burns" was a Negro youth with him during the war, and was used as a body servant. The General did not own the Negro. He came from over on the James River and when McCausland died, Burns was living

Never before published photo of Gen. John McCausland circa 1880. Taken in Cincinnati, Ohio.

at Charleston, West Virginia. General McCausland never owned slaves.

9. General McCausland was a great bee man and liked honey. His favorite breakfast consisted of buckwheat cakes and honey. He would never say grace at the table but would ask his guest to return thanks, if he chanced to know the person was religiously inclined. He liked to help a church and was liberal to a limited extent. While he came of a line of Presbyterians, he liked the Methodist Church because his Aunt Jane Smith had raised him and she was of that persuasion. In spite of his background, when it came to religion the General wasn't anything. However, he was honest and always said that, "a man's [spoken] word should be as good as his written word."

10. As to his personal habits, General McCausland smoked a pipe and cigars. He chewed some, but not much. When he discovered his pipe was making a calloused place on his lip, he quit smoking it. Fond of an occasional toddy, the General did not care for wine or liquor as a beverage. On the whole, the man liked a plain diet.

11. Daily the General arose early. Seldom did he sleep over five hours a night. In the habit of reading, he remained up late at night. His was a sizable library, and his favorite reading was *Official Records of the War*

of the Rebellion, other books on the Civil War, and the *Saturday Evening Post*. When he arose in the morning, he always built the fire in the kitchen stove and rattled the stove a lot so his daughter Charlotte would hear and get up. He called Charlotte "Sissy" as a pet name. Since he did not like novels, he never read them.

12. General McCausland built the first telephone line up the Kanawha River. His farms were connected by it. It extended from Henderson to Buffalo and to Pliny on the south side of the river. His telephone ring was "one long."

13. General McCausland was a great person for a gentleman's agreement. He disliked red tape. He taught his children the old *McGuffey Reader* story of the man and the bundle of sticks. He sought to get his children to stick together and the old story was used to illustrate the advantage of close cooperation.

14. Five feet, eleven inches was the General's height. He lived to be ninety.

15. Among the characters of history that this man thoroughly disliked was President Davis. This was because Davis blocked his promotion to General rank for over a year. Nor did McCausland like General Bradley Johnson who was with him when he burned Chambersburg. On the other hand, the General was fond of General Joseph E. Johnston. When they would visit at the Old White, General Johnston made a pet out of Charlotte McCausland, the General's daughter. General Breckinridge was also a favorite of his.

16. General McCausland liked a story he could laugh at. His was a good sense of humor. He had no use for small talk. While he liked to talk, he always said he liked to talk about something. Not knowing one card from another, this man had no use for gambling or for drinking people. His thinking was clean and off-color stories were foreign to him.

17. Once General McCausland took two of his boys to a Confederate Reunion in Richmond just to show them, as he put it, "the kind of people I used to run around with." He never turned a Confederate soldier from his door without helping him. He lived too far from the old Confederate soldiers. They mainly lived in Virginia and General McCausland's men never got to visit each other very much. Everyone especially his neighbors, addressed him as "General."

18. The General built the stone, castle-like structure that housed him until he died in 1927; the stone was quarried back on the hill from where his home was erected. The stone was hauled to the spot by oxen. Since Greenbrier County, by reason of its Confederate antecedents, had a great place in the General's heart, he, obtained the timber used in building his home from that county. When General McCausland died, his room was left as it was when he passed to Valhalla. A religious item on the walls of his room heralds the fact that "God is Love." He kept a picture of his wife hanging on the wall where it was always in his plain view. He died sitting in his chair. His death was attributed to a slight stroke and an aggravated heart condition.

19. In the evening time of his life, General McCausland often reminisced of his days at the Virginia Military Institute. He talked of "Old Jack," his designation of "Stonewall" Jackson. McCausland was at the hanging of John Brown and said he "could have reached out and touched him." McCausland was there at the execution with the cadets as Squad Marshal. He loved books and recalled that when Hunter burned VMI there were in the Institute library four copies of Audubon's books—the original edition of each. When she was a little girl and the apple of his eye, General McCausland read to his daughter,

General McCausland's grave.

Charlotte, the writings of Byron and Scott, thus forming "Miss Charlotte's taste" for good literature.

20. When his sons were of military age, the dress uniform of the U. S. Army soldier was blue. General McCausland did not want his sons to be soldiers, adding that "I would rather see them dead than wearing a blue uniform." He took great interest in World War I. After reading of the issue of the Battle of the Marne General McCausland gave his estimate of Germany's situation by throwing down his newspaper and exclaiming, "She's whipped!"

21. When the redoubtable old Confederate veteran died in January, 1927, they buried him in his dress suit that he had made in England and which he wore at his wedding. It had been kept in a cedar chest against that inevitable day of his death. The Charleston Chapter of the United Daughters of the Confederacy sent down to the General's home a silk Confederate flag to cover his casket. There was a flood in the Great Kanawha when the General died and they carried his casket down stream to Point Pleasant on an oil burner boat with the Confederate flag secured to the mast. His son, Sam McCausland, who died in April 1953, was the Guard of Honor by the remains of his illustrious father. Part of the way to the General's last resting place, his remains were carried in a hand car over the railroad.

In retaliation of the depredations committed by Major-General Hunter, during his recent raid into Virginia, it is ordered that the citizens of Chambersburg pay to the Confederate States by General McCausland the sum of $100,000 in gold; or in lieu thereof $500,000 in greenbacks or national currency was required to ransom the town, otherwise the town would be laid in ashes in three hours.

The order was signed by General Early.

You'll see . . . what my orders were. That ought to answer the charge that in burning their town I was wreaking my own private vengeance. Good God! I had nothing against their town! My orders were definite and final: I was to go to Chambersburg and demand a ransom. Failing to get the money, I was to burn the town. We needed food, clothes, shoes and forage for the horses—all the things that the Federal soldiers had been helping themselves to. And I still believe it was Early's idea that part of the ransom money . . . was to be used to buy supplies for our men and part to reimburse the citizens whose homes had been destroyed by Hunter. You see . . . I was instructed to give the Chambersburg people a chance to raise the ransom money. . . . Tell me, did you ever hear of Sherman or Hunter or Sheridan giving our people a chance to ransom their homes before burning them?

As a professional soldier, Early's idea did not appeal to me. Left to myself I should have followed General Lee's example. . . . You recall Lee's army molested no private property. Lee issued strict orders against it. But for all that, I could see some justice of Early's demand . . . Frederick

and Hagerstown . . . had paid him ransoms . . . and he had every reason to believe that Chambersburg would do the same. And there was another thing: Early believed that if the North got a sample of what their own armies were doing . . . the result might be a let-up in the destruction of private property in our country.

From David L. Phillips' *Tiger John: The Rebel Who Burned Chambersburg*, Leesburg, Va., 1993.

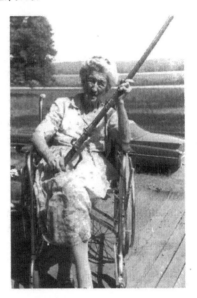

Charlotte McCausland (1884–1971), the General's daughter holds the sword that the grateful citizens of Lynchburg, Va., presented her father. "Miss Charlotte" once spoke about sitting on Gen. Joseph E. Johnston's lap, when as a small girl she attended, with her father, the reunion of Confederate officers held at White Sulphur Springs. BILL WINTZ COLLECTION

General McCausland in later life. WVSA

- 162 -

1925 newsreel photos of General McCausland at his home, "Grape Hill"

A small, frail old man standing on the front porch.

The picture framed in that doorway was nothing like the fierce fighting man I had expected to see. Nobody would have believed that bent figure was the dashing warrior who had won a place in the firmament as one of the galaxy of the greatest shooting stars of their age. It seemed incredible that this courtly, white-haired old gentleman could be the pillar of fire who more than half a century ago had swept through Virginia like a blazing scourge, dealing swift retribution to all who opposed him.

Holding the Lynchburg sword and silver spur.

Nobody would have thought this was the dashing cavalry leader who had swept at the head of his troop into the forts of Georgetown, at the very portals of the enemy's capital; who saved Lynchburg when its doom seemed sealed; who put the torch to Chambersburg, and who more than once tinctured the placid, blue stream on the banks of which he now faces life's sunset, with the blood of his foes.

He personally supervised and directed all the work on the farm, riding his favorite pony which, like its master, refuses to grow old in spite of its accumulating years. The faithful beast, now 29 years old, roams the fields at its pleasure, but never fails to pay a daily visit to the house.

And so one of the last of a constellation of epic heroes lives as he has fought—doing that which needs to be done as well as it can be done—and a little better if possible—always with his face to the front, with head erect and vision unclouded. So he lives, and so he expects to die.

I found the general as gracious host as a soldier; as brilliant a conversationalist as a tactician. Crossing his threshold one is greeted with the hospitality that made Virginia famous. The house, though richly furnished, is Spartan in its simplicity. One is impressed with the military atmosphere. He feels that he is in the general headquarters of an invisible army. But through it all there is an air that puts the visitor at his ease.

Confederate Gen. John McCausland's home in Mason County on Route 35. The general built the house in 1885 and lived here until his death in 1927.

Buffalo Academy was built in 1849 in Buffalo on state Route 62. Gen. John McCausland attended the school, and during the war it served as a Confederate barracks and hospital. The site was known as Camp Buffalo in the early days of the 1861 Kanawha Valley Campaign. The Buffalo Presbyterian Church, adjacent to the Buffalo Academy, was undoubtedly used during the war, also.

Gen. John McCausland's Derringer: The Weapon of Choice

by ED GRANT

Countless weapons were used during the Civil War. Samuel Colt produced thousands of handguns that saw action on both sides. The Springfield Armory was buzzing during the war years, producing more than 525,000 rifled muskets in 1863 alone. While many government issues were distributed among the soldiers, it was not uncommon that soldiers of all ranks carried weapons from home. Generally these were handguns, and most often carried as an extra measure of caution in the event that all else failed.

Confederate Gen. John McCausland carried such weapons all during the war. Most generally noted as the man who burned Chambersburg, Pa., on July 30, 1864, McCausland was known to have carried a pair of Philadelphia Derringers throughout the war. The pair owned by McCausland were the small pocket pistols of .41 caliber. They were identical to the one John Wilkes Booth used that tragic night in Ford's Theatre when President Lincoln was assassinated.

Derringer pistols gained widespread fame and use during the California gold rush. Prior to that time, Henry Derringer, who began making guns in the late 1830s, enjoyed the increasing popularity of his pocket pistols, which were usually sold in pairs. Dur-

ing the war, the popularity fell off considerably. The Derringer was basically effective only at short range. Yet, McCausland carried a pair during the entire war. One reason was obvious. These pistols were very small, and could easily be tucked away in any pocket. At close range, in hand-to-hand combat, these pistols were just as effective as any made by Colt or the dozens of other contractors.

One of the Derringer pistols still resides in the family. The other was given to Confederate General, Nathan Bedford Forrest.

Samuel Hanna, grandfather of General McCausland's wife. He was a resident of Charleston and was associated with the Bank of Virginia. This is a rare miniature (actual size) portrait. They were painted on ivory or ceramic and were usually worn as a locket. They were popular before 1829 when photography first came to America.

George McCausland, grandson of Gen. John McCausland, lives not far from "Grape Hill" and recalls the general, who died when George was seven.

ALL PHOTOS COURTESY GEORGE McCAUSLAND

BENJAMIN A. COLONNA,
Consulting Civil Engineer
WYATT BUILDING,
Washington, D.C.
PHONE MAIN 133

April 11, 1901.

Mr. Henry W. Goodwin,
Charleston, West Va.

Dear sir: Yours of the 5th is received by way of Norfolk, I am sincerely grieved to learn of the death of my old friend Levi Welch; one of nature's noblemen.

Levi and I became classmates in 1862 at the V. M. I, and we're soon fast friends, we led the cadet life until early in the spring of that year when we were ordered to join the Stonewall Brigade at Staunton. The Corps then consisted of about 250 men. Here I first saw Levi in actual service, he was a veteran soldier before he was a Cadet, and I think he was appointed in recognition of gallant and meritorious service in the field. Our campaign ended after the enemy had been driven out of McDowell and beyond Hardy Co. and we were ordered back to barracks at Lexington. The academic year ended as usual on July 4th and the cadets were put in the usual encampment. The monotony of this life was too much for Levi's patience and he secured a leave of absence and joined a Confed. force at White Sulphur Springs just in time to take part in a hot fight that took place in the Hot Spr. Val. and in which action he distinguished himself, he was very highly commended for this in the Newspapers as I well remember having read at the time. On his return to the Corps he was received with acclamation by the Cadets with whom he was always very popular. During the following winter it was necessary to send a reconnaissance party to observe the movement of Gen. Averill, and Prof. Col. Thos. Williamson was ordered to perform that duty. He had the selection of such Cadets as he wished to aid him in the work and he selected Cadets Shriver and Welch and none other, they proceeded on foot in the snow which at the time was thick on the ground and did such good work that Averill's raid made not long after was entirely foiled, and he did not succeed in destroying the salt works as he intended our next and last campaign together began in May 1864 during which we fought the battle of New Market, Va. then went to Richmond and occupied some trenches during the second Cold Har. and thence to Lexington where we joined McCausland's Brig. and witnessed the entrance of the Fed army into Lexington. Thence we went via. Balcony Falls to Lynchburg, where in the trenches we held Genl. Hunter in check until the arrival of Earlys Corps, when we drove Hunter away; thence we went back to Lynchburg and saw the Barracks in ruins. Our class was then graduated. Our last roll call was in front of the central building of Washington College after which the cadets were furloughed and the old Corps of Cadets as it fought at New Market was a thing of the past. In a month or two the Corps was called together against Richmond but it contained only a remnant of the New Market men. After our last roll call as stated, I parted with Levi, we were shortly commissioned Lieutenants in the provisional army for the Con. States but we did not meet again until long after the war.

It does not matter where Levi served, he was one of the truest and bravest of men. He is the sixth out of a class of thirteen and I am admonished that my life is so far spent that I have as much to look forward to beyond the river as I have to turn to here.

I have forwarded your letter to Capt Shafer at Frederick, Md. He was a class-mate of Levi also.

I do not know whether Levi preserved any of his writings or not, but his verses were well received and he was capable of making a reputation, but addicted to doing a good piece of work and reading it to one or two friends and then tearing it up.

I am glad if I have in any manner served you in this matter and shall be pleased to serve you further if I can.

With kind regards, I am yours respectfully,
B. A. Colonna

Note: Levi Welch is buried in the "old circle" of Spring Hill Cemetery, Charleston, near his brother Lt. James Clark Welch C.S.A. who was killed July 17, 1861, at the Battle of Scary Creek. After 93 years a stone was finally placed on Levi's grave in 1994 by R. A. Andre.

A Confederate Veteran Gives a Readable Letter to His Comrades of the "Lost Cause," and wishes to Be Remembered and Not Forgotten

Thursday, November 2, 1893

Editor *Kanawha Democrat*

COMRADE—Seeing several letters lately from my old comrades in-arms of the late war, I thought it might interest some of them to know that I am a survivor of that terrible struggle in which so many of our fellow soldiers fell victims.

I was a volunteer in the 22nd Va. and my first introduction to battle was at Scary, the battle terminating in our victory. I was one of four men, who crossed the creek and captured Colonel Norton and his writer. After the battle of Scary we fell back to Nicholas County, as we were outnumbered by the enemy, and there we had an engagement with Colonel Tyler, which resulted in our victory, and which greatly elated me with my beginning life as a solider—little thinking that my patriotic pride was soon to be humiliated.

We were next ordered to reinforce Wise at Hawk's Nest, where I was left as hospital nurse; my regiment going back to reinforce General Floyd against General Rosecrans at Carnifax. Wise fell back. As a result, I was left between my command and the enemy. As soon as I received this intelligence, I pressed a wagon and started with the sick and wounded to reach our lines. We advanced about four miles, when we were overtaken and captured by Cox's men. That event occurred upon the 28th anniversary of my birthday—September 13, 1861. We were marched to Camp Gauley, where we were held for several days and the number of prisoners increasing to 37. We were then arranged in couples, secured by a rope with cross lines, by which our hands were tied. A giant specimen of Tennessee worked in the lead. We were, indeed, a very much humiliated looking team as we marched in this style to Camp Inzard, where we boarded a steamer for Cincinnati. Arriving there we were marched to the city calaboose and locked in. We were removed at a late hour in the night to the barracks where we were held over Sunday. Being the first prisoners taken to the city we created considerable excitement. We had visitors by the thousands who gazed on us with contempt and some of the women especially, were much disappointed at not seeing us with horns, as they had conceived the idea that rebels had great hooking extremities. At 10 o'clock Sunday night, they started with us again, but our progress was impeded by a mob of negroes and white citizens, who met us at the gate and demanded us of the guards for the purpose of summary dealings. We were put back in the barracks until the guards were reinforced, when they started with us again and they met with the same furious mob, but more intent on taking us as their prey. The third attempt was made before the guards succeeded in getting us away.

Our journey finally terminated with an introduction into Camp Chase prison, where we were held 11 months and 28 days—after which we were taken to Vicksburg and exchanged. While in Camp Chase prison I experienced the excruciating exposure and hardship incident to the soldier's life. Being an extremely wet and disagreeable winter we suffered intensely from cold, as we were minus of stoves and fire until Christmas.

While in prison I contracted a disease that has been detrimental to my health down to the present time. I am not entirely unable to perform labor or attend to business and have been in this condition for eight years. Yet while I am conscious of the fact that my imprisonment has been the cause of untold misery and suffering and will result in a permanent death. It does not weaken my patriot pride and enthusiasm for the cause I justly tried to defend.

Now as I am intruding on space, I will conclude by requesting any of my old comrades-in-arms that may see this to write me a letter, as my pleasure and encouragement now depends upon the sympathy of friends. A letter from one of my comrades would surely be a welcome missive.

Yours respectfully
Isaac Miller
Winifrede, W.Va

Letter courtesy Gerald Ratliff

Unknown Soldiers

The last shot of the American Civil War was fired nearly 130 years ago, but the mysteries created by that terrible conflict continue to surface.

The armies of 1861–1865 had no formal system of identification for the slain. The metal dog tags of the 20th century were unknown and the best hope of a soldier was simply to pin a name tag on his uniform.

In the confusion and haste of battle many men were quickly buried—often where they died and the only notation of their fate might have been a few terse words in their unit records. "Lost in the Kanawha campaign Sept. 1862" at least would tell a grieving family of their relative's death, but they were left without a grave to cover with flowers.

Early in 1993 while going through the old maps and records of Spring Hill Cemetery in Charleston, a local historian discovered a cryptic notation on the marking of a turn-of-the-century map.

It said: "Soldiers 1861–1865" plots 26 through 37 "old circle."

With this tantalizing clue Richard Andre set about the task of trying to find the identity of these long-lost men.

A thorough search of all the old interment books disclosed that indeed the names and even the army of the soldiers was *unknown*.

How had the tragic warriors of the Blue or Gray found their last resting place on this beautiful hilltop overlooking Charleston?

At the outbreak of the Civil War, Charleston and the Kanawha Valley region all the way to the Ohio River was a part of old Virginia and certainly Confederate territory.

In the summer of the first year of the war, overwhelming Federal forces under Gen. Jacob Cox occupied Charleston after a brief battle at Scary Creek on the Kanawha River across from present-day Nitro.

Thus from the last summer of 1861 until September 1862 Charleston was under continuous Federal occupation and subject to military law.

The first Confederate soldier to die in the Kanawha Valley, Lt. James Clark Welch, lost his life at Scary Creek and he is buried with a prominent marker in Spring Hill Cemetery in the same "old circle" section where the "unknown" graves are located.

Among the "old circle" graves are numerous veterans of the 22nd Virginia Volunteer Infantry C.S.A. "The Kanawha Riflemen," including Lt. Welch.

On Sept. 13, 1862, while the great Battle of Antietam raged near Sharpsburg, Maryland—the Confederate army under Gen. William Loring swept down the Kanawha Valley from the Pearisburg, New River area.

Gen. Robert E. Lee, upon learning that about half the Federal garrison in the Kanawha region had been sent to join McClellan's Union forces in Maryland, ordered Loring to retake Charleston and its crucial salt works.

On the morning of the 13th a hot late summer

The dedication service was held in the rain on Sept. 17, 1994.

day the 22nd Virginia, 36th Virginia, 50th Virginia, 45th Virginia, 51st Virginia, 63rd Virginia, 23rd-26th Virginia and 30th Virginia started driving in the blue uniformed pickets near where the eastern limits of Charleston are today.

The records tell us that four federal soldiers lay dead in the streets and fields of Charleston. Six confederates also died in the brief but violent action.

Are the unknown soldiers of Spring Hill those fallen warriors? Convincing arguments can be made for the probability that they are indeed soldiers who were slain on the 13th of September 1862, but were they Union or Confederate?

Bear in mind that the battle was won by the Confederate force as night fell on the 13th. Charleston was firmly in Confederate control. It is hardly likely that the fallen Confederates would have been unknown to their comrades who would have had the task of burying them.

It is difficult to imagine that their names and units would not have been recorded somewhere in the Spring Hill records. Of course, it is possible the records could have been lost over the years. A 1900 Charleston newspaper Memorial Day article refers to the "unknown soldiers" in Spring Hill—so we know that 38 years after the battle their identity was unknown.

So on the one hand is a solid argument that they were not Confederates—but the haunting question looms: Why if they were Union soldiers would they have been buried among the pioneer families of Charleston most of whom were Southern sympathizers?

This point is compounded when we observe that there were half a dozen or so other cemeteries in Charleston at the time, including the Ruffner graveyard that was not far from the thick of battle.

Then there is one last possibility and piece of evidence. In 1925 Thomas E. Jefferies wrote that some years after the Civil War: "During the battle several soldiers were buried where they fell and in later years as streets were extended they were dug up and souvenirs were found!" This seems to fit the puzzle very well, but of course it still does not tell us if they were Union or Confederate or perhaps both.

Historians have pondered this mystery as the 20th century draws to a close, lacking some future discovery of documentary evidence we must leave the story of the "unknown soldiers of Spring Hill" shrouded in the shadows of time.

On Sept. 17, 1994, a memorial stone was set on their gravesite with the following inscription: "Here rest ten or more soldiers known but to God who died in the Civil War, 1861–1865."

Spring Hill Cemetery, Charleston, West Virginia

This Confederate plot, like hundreds throughout the South, was dedicated for the burial of old Southern soldiers—often indigent or without a family plot. This site was purchased by the local United Confederate Veterans near the turn of the century. It is worth noting that none of these graves are battle deaths—simply old soldiers who needed a place to lie down.

This prominent marble needle tells a tale of woe. Both of these soldiers apparently relatives died in Federal prison camps. A very interesting sidelight to this monument is that its virtual twin stands in an ancient graveplot just west of the C&P Telephone building on Charleston's southside. It is also engraved, Estell, and it is perhaps Huston Estell's parents. No doubt their home was nearby (note the graveyard marked on the 1850 C&O map near the front of this book).

Dr. Spicer Patrick was a delegate to the Virginia convention that decided on secession. He voted to stay in the Union and returned to quietly live out the war in Charleston. He was a pioneer in the salt industry.

Dr. Alfred Spicer Patrick was the regimental surgeon of the 22nd Virginia Infantry and the son of prominent Charlestonian, Dr. Spicer Patrick. He rests near his father.

On Memorial Day long ago these Civil War veterans' graves were strewn with flowers—now they are seldom noted. Here we see the monument to William Fife, prominent in the 36th Virginia. Fife was killed along with many others in a train wreck near the present site of Dunbar on July 4, 1891.

One of the most interesting Civil War associated graves in Spring Hill Cemetery is the impressive huge granite slab of the Swann Brothers—all three Confederate officers—John, Thomas, George.

Tom Farley was so proud of his membership in the United Confederate Veterans that he had it noted on his gravestone. It is impressive when you consider that a person's gravestone is their last chance to make a statement to the world.

There are a few Federal soldiers in Spring Hill Cemetery. Edward Wilber rests not too far from his commanding officer, Colonel Polsley.

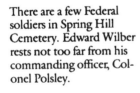

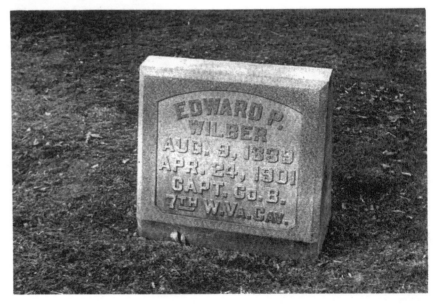

Grave of John P. Hale in Spring Hill Cemetery, Charleston, West Virginia.

John P. Hale in later life.

The story of Adam Brown Dickinson Littlepage's death only involves war in a secondary way. Apparently while his Charleston farm was occupied by Confederate forces in 1861, one of his fine horses was stolen. While in army service in Dublin, Va., he spotted the horse and a duel was fought with the supposed thief. Littlepage was killed. A strange fate for a soldier in wartime. His gravestone is in Spring Hill Cemetery, Charleston.

ADAM BROWN DICKINSON
LITTLEPAGE
DIED IN CONFEDERATE
ARMY APRIL 1862
AT DUBLIN, VA.

REBECC
A. B
M,
FEB

Gen. Albert Gallatin Jenkins' grave monument.

Monument inscribed: *In memory of the gallant Confederate veterans who were members of Camp Garnet-United Confederate Veterans of Huntington West Virginia.*

Spring Hill Cemetery, Huntington, West Virginia

The Confederate section is the last resting place of Gen. Albert Gallatin Jenkins.

Woodland Cemetery,
Ironton, Ohio

As the Civil War became an awful reality to the people of Charleston with the death of C.S.A. Lt. James Clark Welch, so also a scant 70 miles away the citizens of the Ohio River village of Ironton laid to rest Richard Lambert in Woodland Cemetery. Jimmy Welch and Richard Lambert could easily have been friends, but fate decreed that one would wear blue and the other gray and that both would die on that stifling hot day at a place called Scary.

Pvt. Richard Lambert was in Capt. John S. George's Independent Company Ohio Volunteer Cavalry. Above is his grave and family monument.

Capt. Tom Huddleston, Kanawha Rangers. Spring Hill Cemetery, Huntington.

The funeral of Richard Lambert, of the Lawrence County Cavalry, killed at the Battle of Scarey, Western Virginia, July 17th, took place in Ironton, Thursday of last week—Aug. 8—his remains having been brought here on the evening previous. He was buried with the honors of war, the Ironton Rifles, Zouaves, German Guards, and a large number of returned Volunteers joining in the procession; the funeral services at the Grove, prayer by Rev. T. S. Reeve, sermon by Rev. J. M. Kelley. He was about 26 years of age, and leaves a wife and child.

"They who for their country die,
Shall fill an honor'd grave;
For glory lights the soldier's tomb,
And beauty weeps the brave."

THE BANNER

MORY M. AULTZ, Editor and Proprietor

SATURDAY, JUNE 3, 1882.

DECORATION DAY, 1882

The decoration of soldiers graves in Spring Hill Cemetery, was attended by a very large assemblage of our citizens, and a pleasant feature of the observance was, that both classes of our people were present, participating in the sad solemn rite. There was a very numerous attendance of both, a majority perhaps of ladies, always on hand to perform a part in anything that is noble and benevolent. The Grand Army of the Republic, however, was not so largely represented as we had reason to suppose it would be; from what cause we do not know. The procession moved from Kanawha street, under its marshals as previously arranged, led by the Coalburgh Brass Band. Major Snyder's Cadets of the Kanawha Military Institute, (about 40 in line) accompanied the procession under his command, and contributed largely to the occasion, and, but for that organization there could not have been any military salutes.

The grave of the late Col. Polsley, an officer in the late war was first visited. It was profusely decorated with the U.S. flag and beautiful flowers. The G.A.R. It was formed around the sepulcher, together with the cadets. After a prayer and some preliminary remarks by the Marshal Judge J. H. Brown, the Orator of the day, delivered an eloquent patriotic address, at its close a salute was fired by the Cadets, and some further addresses made. The procession to the grave of the late Col. Alexander Quarrier, whose death occurred 65 years ago, supposed to be the only soldier of the Revolutionary war, whose remains repose in the Cemetery. Judge Brown announced the fact in some appropriate remarks; after which a salute was fired; the grave had been previously strewed with

flowers. The next graves visited were those of the late Richard Q. Laidley, a Captain in the Confederate Army, and the late James C. Welch, a young soldier of the same army, who fell at the battle of Scary, July 17, 1861, said to have been the first Confederate soldier killed in battle in the Valley. These graves were strewn with flowers, and salutes were fired. There was a very general decoration of graves in the grounds, as well those of the soldiers as of others, and the floral display in bouquets, wreaths, crosses and other emblems in good taste and profusion.

The address of Judge Brown, was in every way suitable and appropriate to the solemn observance, and is very highly commended.

There was no manifestations of partizanship [sic] on the part of any in the proceedings. The Marshals and other officers in charge of the programme rendering their services to all alike. Maj. Du Bois, Dr. Mayer, Capt. Cracraft, Mr. Noyes Burlew, and Mr. Thomas Swinburn and others, were noticably [sic] conspicuous in distributing flowers, and in endeavors to contribute to the pleasures of the day.

Thus honoring the memory of the dead is a noble attribute of our nature, and we trust that every succeeding year may be an improvement in this observance until it shall be made perfect.

Spring Hill Cemetery has never before looked more attractive as a repose of our dead, than on this solemn occasion.

Let us be zealous and true in our devotion to the memory of the virtuous and worthy dead.

Civil War Historic Sites

McFarland House, 1310 Kanawha Boulevard, Charleston. Built in 1836 by Norris Whitteker, it is one of the valley's oldest and most significant homes. Used as a hospital during the war, the house is now a private residence. It received several direct artillery hits in the 1862 campaign.
COURTESY ELIZABETH HUBBARD

Richard Andre is seen holding a six-pound cannonball (solid shot) that struck the McFarland-Ruby house during the Battle of Charleston in September 1862. It has been at the house for 133 years. The six-pound ball was not heavy, but its penetrating power was considerable.

The Holly Grove Mansion at 1710 Kanawha Boulevard is one of the earliest and most important houses in Charleston. It was built in 1815 by Daniel Ruffner, son of Joseph Ruffner. Because of its size and location along the James River and Kanawha Turnpike, the mansion was used as an inn for many years before the Civil War. Some notable visitors included Henry Clay, Sam Houston, John J. Audubon, and in 1832, President Andrew Jackson. The house was originally built as a two-story structure, but in the early 1900s James Nash purchased the home from a Ruffner heir and made extensive interior and exterior modifications—including the addition of the circular entrance and the massive white columns. Today the home is part of the Capitol Complex, which is owned by the state of West Virginia, and it has been placed on the National Register of Historic Places.
COURTESY HISTORIC PRESERVATION UNIT

Glenwood, on Charleston's west side, was built in 1852 in the Greek Revival style for James M. Laidley, a prominent early salt maker. Judge George W. Summers bought the house in 1857. It is one of the best examples in the county of an original pre-Civil War house. The last resident was Lucy Quarrier, a descendant of Judge Summers. The house is now occupied by COGS.

Cedar Grove was built in 1834 by Augustus Ruffner, fourth son of Daniel Ruffner. This substantial brick house at 1506 Kanawha Boulevard remained in the Ruffner family for more than a century.

This building in Gauley Bridge housed Millers Tavern during the Civil War.

Felix Hansford House, at the mouth of Paint Creek, just off state Route 61 at Hansford, is now just a pile of bricks. The house was built in 1825 and armies of both sides passed by during the war. Confederate troops rested here on their retreat from the Kanawha Valley in July 1861.

Cedar Grove, located in the town of Cedar Grove on the Kanawha River, east of Charleston. The house was built by William Tompkins, prominent businessman of the area, in 1844. The brother of his wife, Rachel, was the father of Gen. Ulysses S. Grant. During the war, Federal cavalry rode through Cedar Grove and threatened to burn down the Tompkins' home. Mrs. Tompkins reportedly produced a letter from General Grant promising immunity for her property. The house is still owned by a Tompkins heir.

Civil War Historic Sites

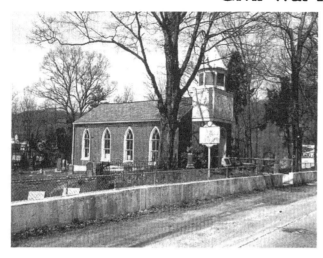

Virginia's Chapel and Slave Cemetery (The Little Brick Church) on US 60 at Cedar Grove. The church was built in 1853 by prominent local businessman William Tompkins at the request of his daughter Virginia. It was built in lieu of a trip abroad as a graduation present. During the war it served as a Confederate hospital and as a stable for Federal cavalry. After the war the Federal Government paid the church $700 for wartime damages. The church has since been restored. The slave cemetery is located behind the church.

The site of Camp White looking west on Charleston's southside. See painting page 149.

University of Charleston, site of Confederate artillery position during the Battle of Charleston.

Orchard Manor housing, on the site of the 1861 Confederate Camp Two-Mile.

Camp Piatt, Federal campsite in the Belle area.

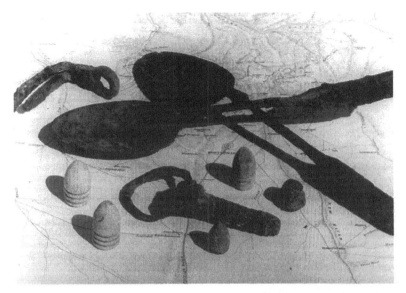

Relics from the collection of J. T. "Slim" Combs, who began searching for Civil War artifacts with a World War II vintage mine sweeper in 1959. He has visited most of the major Civil War battlefields throughout the east. His recreation room is a virtual museum.

These items were found at the Camp Piatt campsite near Belle.

This James projectile was dug up on the southside of the Kanawha River at Gauley Bridge. This is probably a shell that failed to reach the Confederate battery, which was high on the hillside above the bridge. The Confederates had the advantage of firing downward, while Federal cannon had to fire uphill. The keys are for scale.

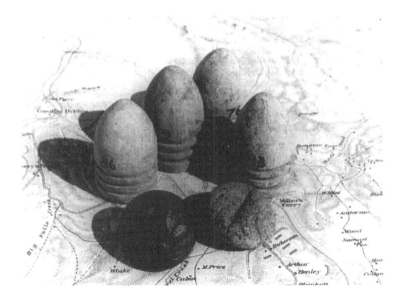

These Minié balls and a Federal uniform button were found near Winfield.

On Charleston's south side, near the foot of the hill next to the carriage trail to "Sunrise," there stands a small stone. In the dim light filtering through the forest, it is easy to forget that automobiles dash by on MacCorkle Avenue a scant few hundred feet away. That cool leafy woodland glen holds a mystery that to this day remains unsolved! Here is what we know: During the construction of Sunrise in 1905, ex-Gov. William MacCorkle unearthed the remains of two women, one blonde, the other a brunette. In shock and astonishment he moved them a few feet to a place next to the carriage trail and went on with the work. Not long after, he talked to Civil War Veteran John Slack, who declared that the women had been accused of spying by the Confederate army and had been executed after a brief "Drumhead" court martial. Since their camp was nearby the women were just taken up the little hollow and buried. The governor accepted this story and had a small stone carved to acknowledge both the event and the burial. Sometime later, another Civil War veteran said the story was true but Slack had the army wrong because the Yankees had shot the women. Finally the governor declared that another Civil War soldier from Lincoln County had confessed on his deathbed that he had been on the Union firing squad and it had haunted him all of his life. So there we have it—two human beings—beloved of someone we know not who, rest near the center of a great city. Were they spies? Did the Federals end their lives or was it the men in gray?

This is a very popular story, but other than MacCorkle digging the bodies up in 1905 and a death bed confession from a delirious old soldier, there has never been any evidence to support it.

The James P. Lanham Memorial Bridge spans the Poca River in Kanawha County. Its namesake joined the Confederate army from Monroe County in 1862. His regiment was the 22nd Virginia under the command of Col. George S. Patton. Lauham was left sick in Maryland in 1864 and was captured and sent to the old Capitol Prison in Washington, D.C. He was released in May 1865 and returned to the Poca area as a farmer and prominent citizen of the county. He died at age 77 and is buried on the hill just above the bridge. The bridge was dedicated on July 25, 1994.

The ferrymaster's house at the end of Ferry Street, just down from the Putnam County Courthouse at Winfield. This ferry site was the first in the area, dating to 1819. The house dates from 1883, but two pre-war tombstones are embedded in the floor. Civil War trenches are reportedly behind the house.

Dutch Hollow Wine Cellars on Dutch Hollow Road in Dunbar is a remnant of the once-thriving wine industry in the Kanawha Valley. Thomas R. Friend operated the business until the Civil War stopped all wine production. After the war, production was attempted again but without success. The cellars have been restored and turned into a city park.

In 1839, Stockton's Inn opened its doors to visiting stagecoaches at Kanawha Falls on the James River and Kanawha Turnpike. The Inn was owned by Col. Aaron Stockton, a slave-owning farmer and coal entrepreneur. The central unit is the original section of the building, with the wings added after the Civil War. Thousands of troops passed by here during the war and in the fall of 1861 Federal forces used it as a quartermaster depot. In November 1861 Confederates fired at the house from the heights of Cotton Hill. The Inn, called The Glen Ferris Inn since 1929, is owned by Elkem Co. and is on the National Register of Historic Places.

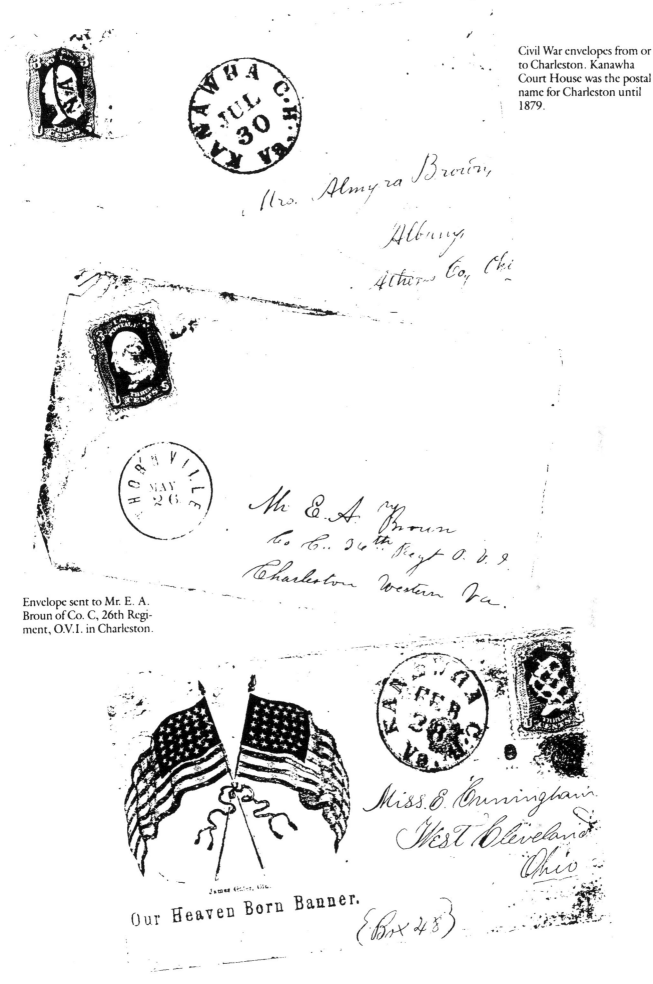

Civil War envelopes from or to Charleston. Kanawha Court House was the postal name for Charleston until 1879.

Mrs. Almyra Broun,

Albany,

Athens Co, Ohio

Envelope sent to Mr. E. A. Broun of Co. C, 26th Regiment, O.V.I. in Charleston.

Mr E. A. Broun
Co. C.. 36th Regt O. V. I.
Charleston Western Va.

James Guley, Cin.

Our Heaven Born Banner.

Miss E. Cunningham.
West Cleveland
Ohio

(Box 48)

The Capitol Statuary

Five statues symbolic of the state's Civil War history are placed around the grounds of the Capitol building.

The oldest statue is that of T. J. "Stonewall" Jackson, famous Confederate general, who was born in Clarksburg and grew up at Jackson's Mill in Lewis County. In 1905 the Charleston Chapter No. 151, United Daughters of the Confederacy petitioned the Senate and House of Delegates for permission to erect a statue of Jackson in memory of soldiers from West Virginia who fought for the Confederacy. Permission was granted on February 25, 1905.

Stonewall Jackson statue

A statue of bronze was decided on and the U.D.C. engaged Sir Moses Ezekiel, well-known artist and sculptor of Rome, Italy, to design and construct it. Five years later the monument was shipped from Rome to Baltimore, Maryland, and thence to Charleston, arriving August 31, 1910.

On Sept. 27, 1910, the statue was dedicated on the grounds of the Capitol, then located downtown. This magnificent statue, weighing over a ton, depicts Jackson in the uniform of a Confederate general with field glasses in his right hand. In his left hand is his drawn sword with the point resting on the ground. The inscription reads:

> Erected as a Memorial to the Confederate Soldiers
> 1861 1865
> By Charleston Chapter
> No. 151
> United Daughters of the Confederacy

A parade was held preceding the unveiling of the monument. It consisted of a parade marshal; the Stonewall Brigade Band of Staunton, Virginia; a battalion of V.M.I. cadets; ex-Governors Atkinson and Dawson; the West Virginia National Guard Band and troops; United Daughters of the Confederacy; the sculptor and many Confederate veterans, including Col. George Imboden of Ansted who led a famous raid through West Virginia in 1863.

An additional stone just underneath the statue reads:

> A Memorial
> to
> Samual S. Green S.G.T.
> Montgomery's Battery
> Cutshaw's Battalion L.A.
> C.S.A.
> Brig. Gen. W.Va. Div. U.C.V.

Mountaineer statue

On July 25, 1926, five years after the downtown Capitol burned, the statue was moved to the grounds of the new Capitol building. In 1976, the Jackson figure was removed along with the Mountaineer and the Union soldier statues, and taken to Kingwood in Preston County for refurbishing. Jackson and the Mountaineer had to be moved to make room for the new Cultural Center. Jackson was reset at its present location on the southeast corner of the Capitol's lawn near the intersection of Kanawha Boulevard and California Avenue. An identical copy of Ezekiel's "Jackson" statue stands on the parade grounds of the Virginia Military Institute in Lexington, Virginia.

This is not the end of the Jackson story however. In 1978, Governor Rockefeller, at the request of Charleston Chapter No. 151, U.D.C., the group who originally erected the statue, ordered Jackson to be turned 90 degrees to look out over the Kanawha River. This was done at a cost of $1,200 so the public could see the front of the statue from either the boulevard or California Avenue.

The Mountaineer statue, which now graces the Capitol lawn at the northeast corner of Washington Street and California Avenue, was unveiled on December 10, 1912, at the downtown Capitol. Col. William Seymour Edwards, speaker of the House of Delegates in the 1890s, was the guiding force for this monument.

Edwards began work to raise funds for his project. The Mountaineer was a symbol of the men of western Virginia who formed themselves into Home Guards and responded to Lincoln's call to arms in 1861. By their actions they helped to save western Virginia for the Union.

In 1907 a contract was made with sculptor Henry K. Bush-Brown to create an 8-foot bronze memorial. Bush-Brown spent time in the state to familiarize himself with the mountain men whom he was to depict. He spent some time in Webster County and met members of the Hamrick family. Stories vary as to whether it was Eli "Rimfire" Hamrick or his 6-foot-6-inch brother, Ellis, who really was the model for the statue. It took the sculptor two years to find the model and three years for actual construction. The 8-foot statue was dedicated by Colonel Edwards, "To the hallowed memories of the brave men and devoted women who saved West Virginia to the Union."

Union statue

The third soldier monument on the present Capitol grounds is that of a Union soldier. In 1927 the state legislature created a Soldiers and Sailors Memorial Commission. This commission was to plan the building, on the Capitol grounds, of a monument in memory of men of the state who fought for the Union. Governor William G. Conley was chairman along with four appointed members, all veterans of the Union Army.

The state set aside $15,000 for the project, in contrast to the other two statues, which were paid for by private funds. A New Martinsville, West Virginia, monument firm was awarded the contract. When the statue arrived and was to be placed at the southwest corner of the Capitol grounds, an objection was raised by the architect, Cass Gilbert. He objected because it did not have artistic lines and was not in keeping with his grand design.

Chairman H. S. White of the memorial commission and the other commissioners were adamant about the location and their decision prevailed. It remains at that location, carrying the inscription:

> "To the 32,000 soldiers, sailors and marines contributed by
> West Virginia to the Union—1861 to 1865."

The statue, "Abraham Lincoln Walks at Midnight," was originally designed by Fred Torrey of Fairmont in the late 1930s and exhibited at the 1939–40 New York World's Fair. It was a 42-inch high plaster statue designed for bronzing. Torrey, who died in 1967, was one of the most noted sculptors of his day. His wife, Mabel, sometimes collaborated with him. Torrey's works stand in many states from Colorado to New York.

Abraham Lincoln statue

It was proposed to place the statue on the south side of the Capitol building facing south so that light would always be on the face. A sum of $30,000 was collected from various sources statewide and a contract given to well-known Charleston artist and sculptor, Bernie Wiepper to reproduce a 9½-foot statue from the 42-inch model. Wiepper developed a novel method for the exact reproduction which had never before been used.

The creation of the state of West Virginia was a direct result of the Civil War. In 1870 the capital was moved from Wheeling to Charleston. This remarkable picture of the first Charleston capitol was recently discovered and is believed to be the only photo of this building in existence. Located on the same site as the later 1885 capitol on the block bounded by Washington, Lee, Capitol and Dickinson streets; it was built largely through the efforts of Dr. John P. Hale, a leading Charleston citizen. The photo was taken about 1873 and was part of a stereopticon series published to commemorate the coming of the Chesapeake and Ohio Railroad in that year. We are indebted to Mr. Julius E. Jones of Richwood for allowing us to copy his original.

This 1873 view of Charleston from South Hills is believed to be the oldest scenic photograph of the city in existence. The "Hale House" is at the left middle of the photo. The 100-room hotel, built by Dr. John P. Hale in 1870, burned in 1884. Until the recent discovery of this photo the only image of the "Hale House" was a pen and ink drawing. The building was distinguished by its French style roof. The "Hale House" was located on the northwest corner of Kanawha Boulevard and Hale Street. In the upper right section of the photo the rear of the 1870 capitol can be clearly seen. Near the middle just to the left of the tree foliage is the steeple of the 1830-vintage Presbyterian church at the northwest corner of Hale and Virginia streets. It is interesting to note the character of the river banks and the low water level. In the top left distance is mostly open farm lands and although this photo was taken eight years after the Civil War it gives a general idea of what the town looked like in the war years.

Epilogue

After Lee's exhausted men finally laid down their arms at Appomattox, what were the lasting results of the Civil War in the Kanawha Valley?

First and foremost the people of western Virginia finally got the statehood many of them had dreamed of for years. No more looking to Richmond.

As with the rest of the South, slavery was a thing of the past, although as a practical matter not much was initially changed for the black population who had scant education and little understanding of true independence.

The shock waves of the Confederate defeat reverberated through the valley as men like Dr. John P. Hale strove to begin anew and pull together the shattered remains of their former prosperous lives.

The return to Charleston in 1865 was not easy for the weary men who had placed all their bets on the Confederacy. Men who had practiced law before the war found themselves without a profession as ex-rebel soldiers were denied many former rights.

In predictable observance of human nature the Unionists went out of their way to make life miserable for former Confederates.

The salt business never returned to its former production largely because of new sources in other areas. But for a few years after the war over a million and a half bushels was shipped annually. From a postwar high of 1,822,430 in 1869—production fell to less than a million in 1875 and 150,000 in 1890. All this compared with the all-time high of 3,224,786 bushels in 1846.

While salt declined, coal brought new prosperity to the fledgling state as did timber and other natural resources.

It is probably safe to say that the Kanawha Valley shook off the tragedy of the Civil War far easier than the deep South. The nearness to the Ohio markets linked by the river finally softened the bitterness of war in all but the hardest hearts. A few like Gen. John McCausland never forgave or forgot and went to their graves unreconstructed.

It is rather ironic to consider that the two men most responsible for assuring the location of the West Virginia State Capital in Charleston were both former Confederate soldiers—Dr. John P. Hale and Senator John Kenna.

There can be no doubt that the Kanawha Valley was forever changed by the Civil War, especially Charleston, which grew to be a thriving metropolis as the capital city and the center of a booming industrial complex.

As we approach the dawn of the 21st century we can look back at the rich tapestry woven from nearly every thread of American history. With confidence born of triumph over tragedy let us go forward with faith that in the great Kanawha Valley—the best is yet to come.

John E. Kenna was one of the most prominent citizens of Kanawha County and led the fight to make Charleston the state capital in 1870. He was born in 1848, served in the Confederate army and was admitted to the bar in 1868. He was elected county prosecutor in 1872, a judge in 1875, U.S. Congressman in 1876, 1878, and 1880, and a U.S. Senator in 1883 and 1889. He died in Washington, D.C., at a young 45 years of age. WVSA

Selected Bibliography

Cohen, Stan, *The Civil War in West Virginia, A Pictorial History*, Pictorial Histories Publishing Co., Charleston, W.V., 1976.

_____, with Richard Andre, *Kanawha County Images, A Bicentennial History, 1788-1988*, Pictorial Histories Publishing Co., Charleston, W.V., 1987.

_____, *A Pictorial Guide to West Virginia's Civil War Sites and Related Information*, Pictorial Histories Publishing Co., Charleston, W.V., 1990.

Dayton, Ruth Woods, *Pioneers and Their Homes on Upper Kanawha*, West Virginia Publishing Co., Charleston, W.V., 1947.

Dickinson, Jack L., *Tattered Uniforms and Bright Bayonets, West Virginia's Confederate Soldiers*, Marshall University Library Associates, Huntington, W.V., 1995.

_____, *Jenkins of Greenbottom: A Civil War Saga*, Pictorial Histories Publishing Co., Charleston, W.V., 1988.

Geiger, Joe, Jr., *Civil War in Cabell County, West Virginia, 1861-1865*, Pictorial Histories Publishing Co., Charleston, W.V., 1991.

Lowry, Terry, *The Battle of Scary Creek, Military Operations in the Kanawha Valley, April–July 1861*, Pictorial Histories Publishing Co., Charleston, W.V., 1982.

_____, *22nd Virginia Infantry*, H.E. Howard Inc., Lynchburg, Va., 1988.

Marshall, Paul D. & Assoc. Inc., *Fort Scammon and A History of the Civil War in Charleston and the Kanawha Valley, West Virginia*, Charleston, 1986.

McKinney, Tim, *The Civil War in Fayette County, West Virginia*, Pictorial Histories Publishing Co., Charleston, W.V., 1988.

Pauley, Michael, *Unreconstructed Rebel, The Life of General John McCausland C.S.A.* Pictorial Histories Publishing Co., Charleston, W.V., 1992.

Phillips, David L., *War Diaries: The 1861 Kanawha Valley Campaign*, Gauley Mount Press, Leesburg, Va., 1990.

_____. *Tiger John: The Rebel Who Burned Chambersburg*, Gauley Mount Press, Leesburg, Va., 1993.

Sutphin, Gerald & Richard Andre, *Sternwheelers on the Great Kanawha*, Pictorial Histories Publishing Co., Charleston, W.V., 1991.

Wintz, Bill, *Civil War Memoirs of Two Rebel Sisters*, Pictorial Histories Publishing Co., Charleston, W.V., 1989.

_____, *Annals of the Great Kanawha*. Pictorial Histories Publishing Co., Charleston, W.V., 1993.

About the Authors

Richard A. Andre

For Richard A. Andre his work in the preservation of history has answered that age-old question: "Why was I born?"

This is the fifth book Andre co-authored on West Virginia history, including the bast-selling *Kanawha County Images, Roar Lions Roar, Capitols of West Virginia* and *Sternwheelers on the Great Kanawha*.

Retired from the mortgage business in 1976, Richard devoted the rest of his life to the study of the past and was especially interested in the preservation of vintage photographs.

Andre's interest in the Civil War was heartfelt since he had a grandfather who wore the blue and a great-great-grandfather who wore the gray!

Richard was born in 1940 and was a lifelong resident of Charleston. He passed away in 2015.

Stan Cohen

Stan Cohen, a Charleston native, is a graduate of Charleston High School and West Virginia University. After working for some years as a geologist, a ski shop owner and director of a historical park, Cohen established Pictorial Histories Publishing Company in 1976. Since then he has authored or co-authored 50 books and published 180. His West Virginia titles include: *The Civil War in West Virginia, A Pictorial History; West Virginia's Civil War Sites; Historic Sites of West Virginia; Historic Springs of the Virginias; King Coal, A Pictorial Heritage; Covered Bridges of West Virginia* and *Kanawha County Images, Roar Lions Roar* and *Capitols of West Virginia* with Richard Andre. He divides his time between Charleston and Missoula, Montana.

William D. Wintz

Bill Wintz, a descendent of pioneer families of the Kanawha Valley, first learned to appreciate his heritage from his grandparents. His longtime interest in the history of the Kanawha Valley and West Virginia kept him involved in regional historical organizations for many years. He was active in the Kanawha Valley Historical and Preservation Society, the Kanawha Valley Genealogical Society, and the West Virginia Historical Society. He was also on the Editorial Advisory Board for the West Virginia History publication.

Wintz was a retired Engineering Technician from Union Carbide Corporation and an infantry veteran of WWII, having participated in the Normandy Landing and the Battle of the Bulge. After retirement he wrote five books including a basic *History of Putnam County, Recollections of Mollie Hansford, Nitro the WWI Boom Town, Civil War Memoirs of Two Rebel Sisters,* and *Annals of the Great Kanawha*.

In 1979, he was awarded a National Commendation from the American Association for State and Local History for his historical contributions. In 1994 Wintz reeived the Virgil A. Lewis award when he was named Historian of the Year by the West Virginia Historical Society.

Wintz passed away in 2013.

Made in the USA
Monee, IL
23 June 2023

36651183R00111